The Railways and Tramways of the Isle of Man

Douglas Horse Tramway No. 1, one of the last products of Milnes, Voss & Co. Ltd before they closed in 1913, stands at the southern terminus, awaiting its return to Derby Castle.

The Railways and Tramways of the Isle of Man

Barry Edwards

OPC

Oxford Publishing Co.

Dedication

To my Grandfather, Norman C. Edwards, who sadly passed away before this book could be completed.

Bibliography

Isle of Man Railway, Ian McNab, Greenlake Publications 1945.
Isle of Man Album, W.J. Wise & J. Joyce, Ian Allan 1968.
Isle of Man Tramways, F.K. Pearson, David & Charles 1970.
Horse Trams The First 100 Years, Harry Constantine, Douglas Corporation 1975.
Cable Tram Days, F.K. Pearson, Douglas Cable Car Group 1977.
Discovering Isle of Man Horse Trams, Harry Constantine, Manxman Publications 1977.
The Isle of Man Railway 4th Edition, J.I.C. Boyd, Oakwood Press 1977
Polar Bear and the Groudle Glen Railway, C.G. Down & D.H. Smith, Brockham Museum Association 1977.
On the Isle of Man Narrow Gauge, J.I.C. Boyd, Bradford Barton 1978.
Manx Electric Railway Album, Dr R. Preston Hendry & R. Powell Hendry, Hillside Publishing 1978.
Manx Northern Railway, Dr R. Preston Hendry & R. Powell Hendry, Hillside Publishing 1980
Manx Electric Railway, Dr R. Preston Hendry & R. Powell Hendry, Isle of Man Railways 1982.
Isle of Man Railway, Dr R. Preston Hendry & R. Powell Hendry, Isle of Man Railways 1983.
British Trams & Tramways, Peter Johnson, Ian Allan 1985.
Manx Experience 5th Edition, G. Kniveton, Manx Experience 1987.
Snaefell Mountain Railway, A.M. Goodwyn, MER Society 1987.
Groudle Glen Railway, D.H. Smith, Plateway Press 1989.
All About the Manx Electric Railway, A.M. Goodwyn, MER Society 1989.
Groudle Glen Railway, Tony Beard, IOMSRSA 1990.
Encyclopaedia of Narrow Gauge Railways of Great Britain and Ireland, Thomas Middlemass, PSL 1991.
Fleet History of the Isle of Man Department of Tourism and Transport, PSV Circle 1991.
The Manx Electric Railway, G. Kniveton, Manx Experience & Isle of Man Railways 1991.
British and Irish Tramway Systems since 1945, M.H. Waller & P. Waller, Ian Allan 1992
Manx Steam Railway News, various issues, IOMSRSA.
Manx Transport Review, various issues, MER Society.

A catalogue record for this book is available from the British Library.

ISBN 0-86093-507-8

Oxford Publishing Co. is part of the
Haynes Publishing Group PLC
Sparkford, near Yeovil, Somerset, BA22 7JJ

Haynes Publications Inc.
861 Lawrence Drive, Newbury Park, California 91320, USA

Printed in Great Britain by Bath Press Ltd.

Typeset in Times Roman Medium
by BPCC Techset Ltd, Exeter

All photographs not otherwise credited were taken by the Author

Frontispiece: Steam Railway No. 11 *Maitland* pulls away from Port Soderick station during its journey from Douglas to Port Erin. This locomotive was built by Beyer, Peacock in 1905.

Title page: Manx Electric Car No. 22 with trailer 40 passes Dreemskerry station with a Ramsey bound service on 17th August 1981. No. 22 was destroyed by fire during 1990 but has since been rebuilt.

Contents

Sea Lion stands in the platform road at Lhen Coan station on the Groudle Glen Railway, awaiting departure for the Headland. Four members of the Isle of Man Steam Railway Supporters Association discuss the day's events.

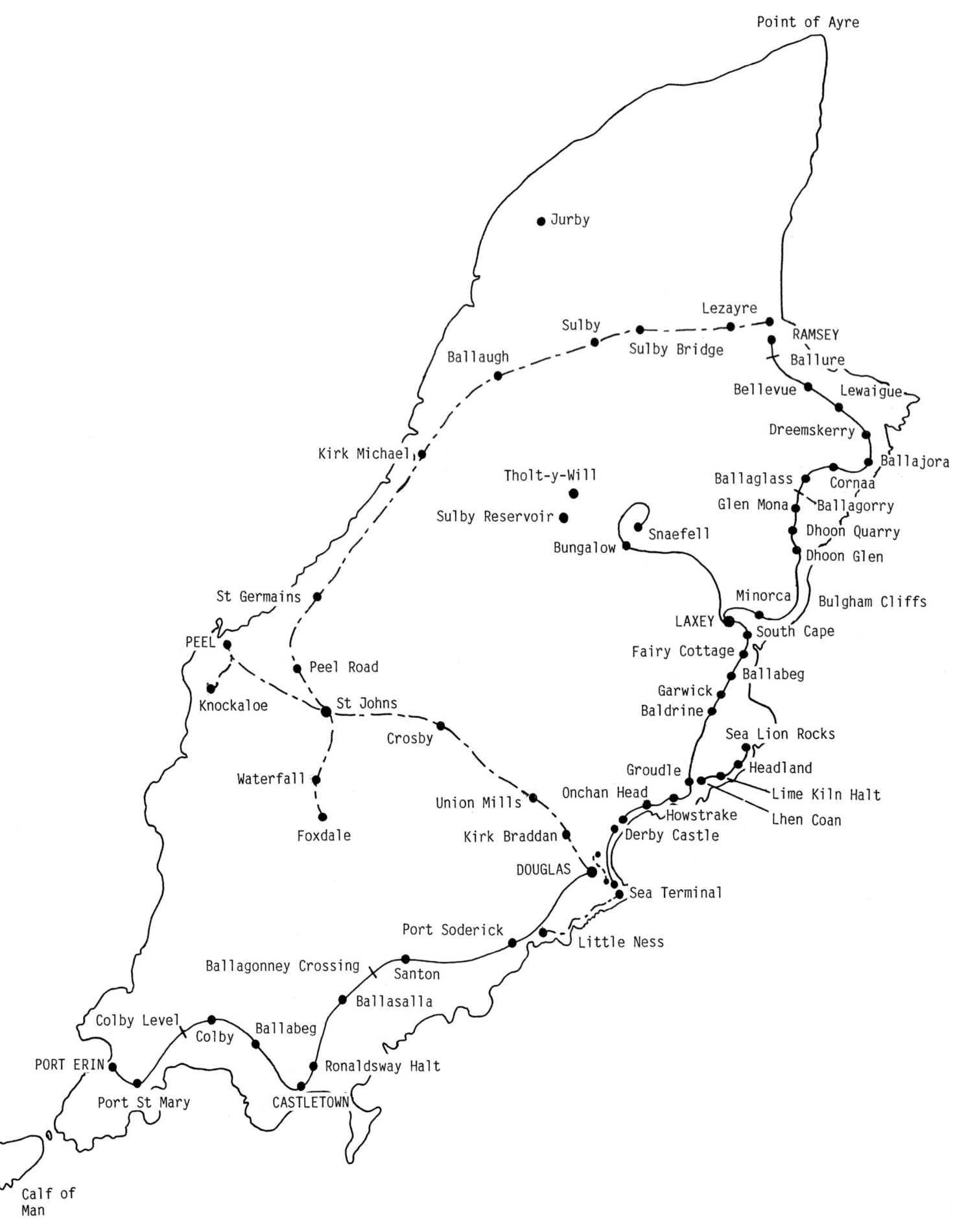

Foreword

by Robert H. Smith, BA, MCIT,
Transport Executive,
Isle of Man Railways.

It was five years ago that I left London Transport to join the very different world of operating the British Isles' largest vintage transport network. Those years have flown by which must prove what a pleasure it is to work on Isle of Man Railways.

The railways and tramways of the Isle of Man offer the enthusiast the unique opportunity to travel behind elegant 3ft gauge steam locomotives, ride on 19th century electric tramcars and a mountain tramway, clip-clop along the Douglas promenades on a horse drawn tram and visit the 2ft gauge Groudle Glen Railway, all in the same day – although I would recommend a stay of longer to make sure you see everything! The Island has five systems with around 40 miles of railway and a combined age of over 530 years.

The Manx Electric Railway has the oldest operational electric tramcar, still working on its original system, in the world. The steam railway is the longest narrow gauge example in Britain. The horse tramway is the only one in the world still in regular use and the Groudle Glen Railway represents one of the finest narrow gauge restorations ever undertaken.

The Island is fortunate in having many skilled staff on its railways who are able to keep the trams and steam locomotives in pristine condition, ready to carry the many thousands of visitors to the Island who will arrive to travel on them.

The Railways and Tramways of the Isle of Man gives the reader a geographical tour of all the remaining operational lines, starting in the north and working south. A short history opens each section of the book and around 240 photographs with informative captions will no doubt entice you to come and experience the real thing.

We look forward to welcoming you aboard!

Robert Smith

When the Snaefell Mountain Railway first opened in 1895, its terminus was a lengthy walk from the Manx Electric station. This picture shows the intermediate (1897) terminus, with a spoil heap from the nearby mine to the right of car No. 2.

Manx Museum and National Trust

Introduction

The railways of the Isle of Man, the youngest of which is now 97 years old, provide a unique collection of working vintage transport. Since their buoyant start and rapid expansion in the late 1800s and, in spite of closures during the two World Wars, they have beaten off complete closure threats. One has come back from virtual extinction, and they have all emerged into the 1990s in better condition, with perhaps a more secure future, than ever before. This book stands as a tribute to all the staff, past and present, who have given their working lives to keeping the railways operational, together with much of the original rolling stock.

Four of the five systems are controlled by the Government Department of Tourism and Transport. The Island has its own Government (Tynwald), elected members taking seats in the House of Keys. New laws only come into force after they have been read in Manx and English from Tynwald Hill at St Johns, once a busy railway junction, on Tynwald Day, 5th July.

I made my first visit to the Isle of Man in 1976, returning three years later armed with plenty of film to begin a collection of photographs, to which I have added nearly every year since. For those interested in photography, the 1979 pictures were taken on a Zenith camera, which was replaced by a Mamiya 645 for the next and all subsequent visits, Kodak Tri-X and, latterly, T-Max film having been used.

The Railways and Tramways of the Isle of Man is an up-to-date photographic survey of the five surviving lines, with brief glances into the past and short histories of each system. Comprehensive fleet lists complete the story. Some items shown in the lists as stored, may technically have been withdrawn, the nature of these railways being that any piece of stock not actually broken up, might be returned to service at any time if required.

No book is possible without the assistance of others and thanks must go to all who have so willingly provided photographs for inclusion in this volume. Thanks are also due to the staff of the Isle of Man Railways, in particular to Robert Smith, Transport Executive, and Alan Corlett. Thanks also to Peter Nicholson of OPC for his encouragement. Special thanks must go to my Father, who's assistance has been invaluable and generous and, to my wife Carol for her support and understanding over the past six months.

It has been a pleasure to compile this volume; my only hope now is that the reader will enjoy this pictorial tour of the Railways and Tramways of the Isle of Man.

Barry Edwards
Ickenham,
Middlesex

Car 5 and trailer 46 load up at Ramsey station, ready to form the 11.30 to Derby Castle on 17th August 1981. The guard, whose foot can be seen on the rear steps of No. 5, is positioning the trolley pole on the overhead wire. The $17\frac{3}{4}$ mile journey to Douglas will take 75 minutes.

The Manx Electric Railway

The first thoughts of a railway from Douglas to Laxey and Ramsey came in 1882 with the formation of the Douglas, Laxey & Ramsey Railway Co., which planned a line leaving the steam railway at Quarterbridge, climbing towards Onchan and on to Laxey. This proposal came to nothing, as did a number of other projects, including an adventurous plan to tunnel through Bank Hill with a line starting near the steam railway terminus in Douglas.

Eventually, in 1892, Alexander Bruce, Manager of Dumbells Bank in Douglas since 1878, and Frederick Saunderson, an engineer who had come from Ireland in 1865 and who had connections with the family of Richard Rowe, Captain of the Laxey Mines, sought and were granted powers by Tynwald to develop Howstrake Estate, an area to the north of Douglas. The Estate was purchased in the name of Mr Saunderson, with backing from a Mr Alfred Lusty, a wealthy London merchant who had retired to the Island and lived on the estate. Douglas Bay Estates Ltd was registered in September

1892 with a capital of £50,000, construction of an electric tramway getting under way as soon as possible, the first stage of what is now known as the Manx Electric Railway.

Electrification was carried out by Mather & Platt of Salford Ironworks, while G.F. Milnes of Birkenhead built the first three tramcars along with six trailers. Two of these original tramcars, Nos 1 and 2, are still in service today. The area now occupied by the depot at Derby Castle was once an inlet of the sea, and was filled in to carry the tramway past, the area behind being big enough to accommodate the depot and power station. The name Derby Castle was that of a private dwelling which was taken over to allow construction of a ballroom and theatre on the site now occupied by the Summerland Leisure Complex.

The line officially opened on 7th September 1893, carrying 20,000 passengers between then and 28th September when services were suspended to allow construc-

Car 2 with Royal trailer 59 waits at Ramsey for its party of passengers from the Isle of Man Railway Society, who are enjoying short rides on the former Ramsey Pier train which stands in front of No. 2. The tram will consume 50 units of electricity on its journey, the trolley wheel rotating 145,000 times. 27th May 1989.

tion to begin of an extension to Laxey, approved by Tynwald on 17th November 1893. The line was doubled throughout its length, the Groudle Viaduct built and a terminus near the present depot at Laxey constructed, opening on 27th July 1894. To accommodate the increase in traffic expected with the extension to Laxey, six more tramcars were delivered from G.F. Milnes during 1894, along with six more trailers, all bearing the name Douglas & Laxey Coast Electric Tramway Co. Ltd on their bodysides, the name to which the company had changed during the construction of the Laxey extension. The company name was changed again, to Isle of Man Tramways & Electric Power Co., when the horse tramway was purchased on 30th April 1894. Control of the horse tramway was taken from 1st May. Part of the deal of purchasing the horse tramway was to construct a cable tramway to serve the upper Douglas area and the Upper Douglas Cable Tramway, as it was known,

opened on 15th August 1896.

Construction of the Snaefell Mountain Railway from Laxey to near the summit of the Island's highest mountain was undertaken in 1895. It opened on 21st August and, after completion at a cost of around £40,000, the line was sold to the Isle of Man Tramways & Electric Power Co. Ltd for £72,500!

The success of the Douglas to Laxey line prompted the company to petition Tynwald for an extension to Ramsey. Although consent was not received until 1st November 1897, it seems likely that some construction preparation had already started. Ballure was reached by July 1898 and Ramsey by July 1899, bringing the total length of the line to $17\frac{3}{4}$ miles (28.5 km). The official opening ceremony of the extension was held on 24th July 1899.

Dumbells Bank collapsed in February 1900, with outstanding loans to the Railway of £150,000, causing the

On the same day as the previous picture, the former Ramsey Pier locomotive, but now with modified 'steam outline' bodywork, propels one of its coaches along the Manx Electric track with Society members aboard.

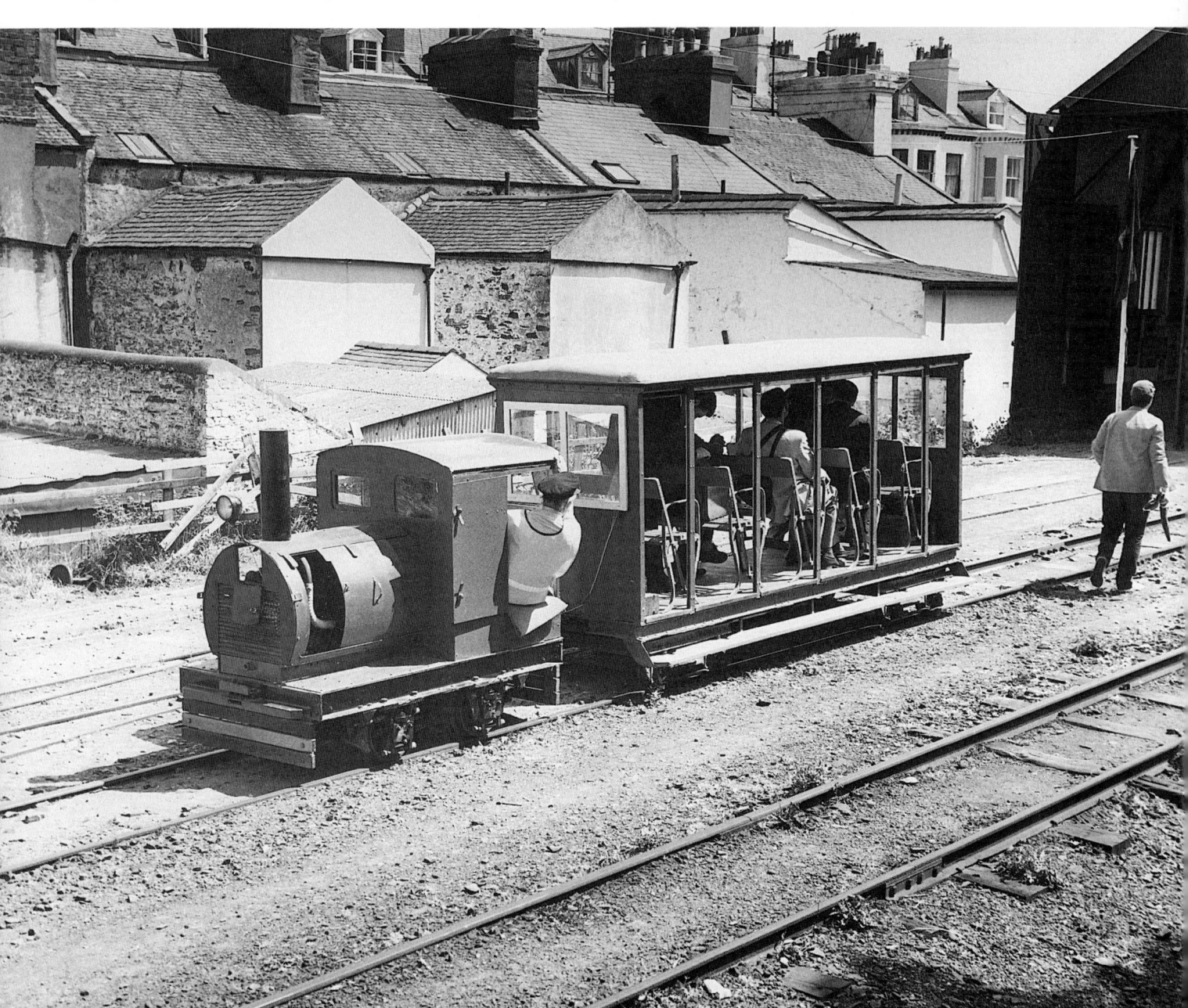

The same locomotive, Hibberd 2027 of 1937, a Y type 'Planet' 4-wheel petrol locomotive when in original condition at the landward end of Ramsey Pier on 29th May 1976.

Peter Nicholson

Isle of Man Tramways & Electric Power Co., and a considerable number of other companies, to go into receivership. A £225,000 offer from the British Electric Traction Co. for the entire undertaking was rejected in 1901. Douglas Corporation offered £50,000 for the horse tramway and cable tramway in late 1901 and this was duly accepted. January 1902 saw an offer accepted from a Manchester based syndicate, backed by a continental merchant banker, for £250,000 for the electric tramway, the final settlement being made in September 1902.

The Manx Electric Railway Company was incorporated in London in November 1902, purchasing the tramway from the Manchester syndicate for £370,000. The financial problems of this period were compounded by the involvement of several senior members of the Isle of Man Tramways & Electric Power Co. in the Blackpool & Fleetwood Tramway Co. Ltd, which built the 8-mile long system there to the mainland standard gauge of 4ft 8½in, much of the rolling stock being similar to that on the Island.

The new company had considerable problems in its early days, with unreliable equipment and a backlog of maintenance. Much re-equipping took place both to the rolling stock and to the permanent way, and by 1906 the line was as up to date as any other.

The Manx Electric soon became a valuable asset for the Island's residents, as well as being a visitor attraction with its increasingly old rolling stock and beautiful scenery. However, during the period 1928 to 1932, a serious accident and three setbacks befell the line.

On 8th August 1928 car No. 1 with trailer No. 39 well laden had stopped in the Fairy Cottage area to collect overhead maintenance staff, when car No. 16 with trailer No. 56, also well laden, came round the curve and were unable to stop, causing a serious rear end collision in which 32 people were injured. An enquiry resulted in the stopping on curves being banned.

The greatest loss to the system occurred on the night of 15th April 1930 when fire broke out in the Laxey depot, destroying motor cars Nos 3, 4, 8 and 24, seven trailer cars, all three tower wagons, a mail van and much other equipment. Only the depot and three trailers were rebuilt. Later in 1930, serious flooding in Laxey after a violent storm was blamed on the Manx Electric's weir which formed part of the company's generating station. A court ruling ordered the company to clear 4–5,000 tons of rubbish from the river bed and pay all the legal costs.

The third disaster occurred on 3rd April 1932, when fire destroyed the hotel and refreshment room at Dhoon Glen, which were never rebuilt.

WALKING
ALONG THIS
LINE IS
STRICTLY
PROHIBITED

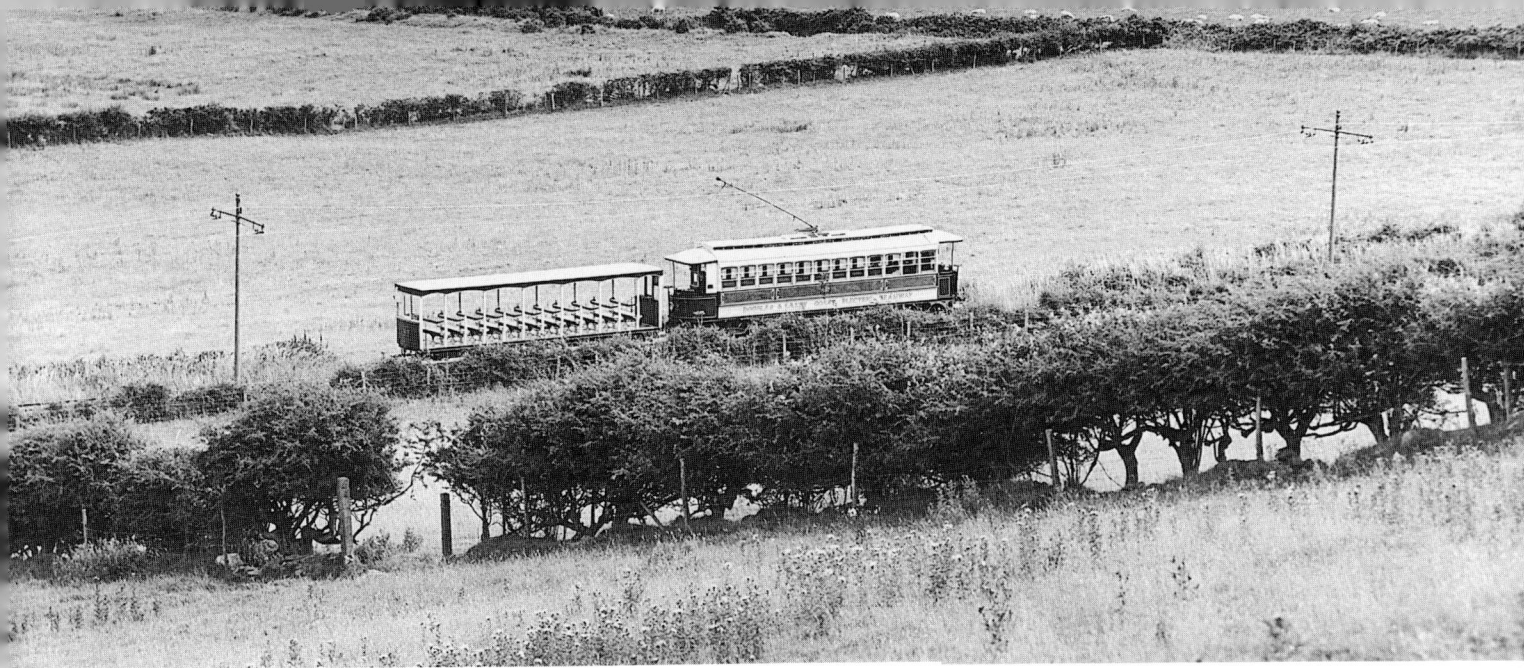

No. 2 with trailer 37 heads towards Derby Castle along the longest straight stretch of the whole line, between Lewaigue and Dreemskerry.

Left: No. 7 crosses Ballure Viaduct on its journey to Derby Castle. The viaduct was built in 1899 by Francis Morton & Company of Garston, Liverpool and spans a gap of 160ft.

Cornaa station shelter, $13\frac{1}{4}$ miles from Derby Castle. This is a request stop, as indeed are many of the stations.

Left: No. 20 with trailer 47 passing Lewaigue station, having completed just under two miles of its journey from Ramsey to Derby Castle.

This rustic shelter at Ballaglass, photographed in 1981, was demolished in 1985 and replaced by a new shelter. Passengers alight here to visit the nearby fish hatchery and 17 acres of fine glen scenery.

Passing the site of Ballaglass power station, now privately owned, is No. 7 with trailer 42 heading north to Ramsey during August 1981.

Right: A view looking north from the Ballagorry bridge shows No. 21 with trailer 47 heading south with the 11.30 Ramsey to Derby Castle, on Sunday, 24th May 1992.

Throughout its entire $17\frac{3}{4}$ mile length the Manx Electric only has two overbridges. The first carries a track over the line at Ballagorry cutting, one side of the bridge housing one of the line's electric sub-stations. No. 2 with trailer 37 heads north towards Ramsey and is about to pass under the bridge.

Glen Mona station building with a postbox at its side. At one time the tram crews used to collect the mail along the line and deliver it to either Ramsey or Derby Castle.

No. 5 and trailer accelerates away from Glen Mona with a Ramsey to Derby Castle service. Glen Mona station is situated by the fourth pole behind the trailer.

No. 7 with trailer 42 passes Dhoon Quarry sidings, whilst working a Ramsey to Derby Castle service during August 1981. The siding holding the wagons remains today but that in the foreground has since been removed. At one time there was a more extensive siding layout serving two quarries, one either side of the line.

Power feeder pillars are situated at intervals along the route. This one, still showing the Isle of Man Tramway & Electric Power Co. initials, was photographed at Dhoon Quarry in 1981.

A close-up of the two wagons shown in the picture above.

Running high above the sea along Bulgham Cliffs is No. 19 with a well-loaded trailer. No shutters are down on the trailer, indicating a warm day. Photographed in May 1991.

Right top: Minorca station is located on the northern side of Laxey Glen. Photographed from the road in 1981, the station building is visible at the top of the steps.

Right below: On the night of 30th September 1990 No. 22 caught fire in the depot at Derby Castle. A replacement car was built on the original chassis and entered service in May 1992. It is seen here on the 24th of that month leaving Laxey with a special to Ramsey.

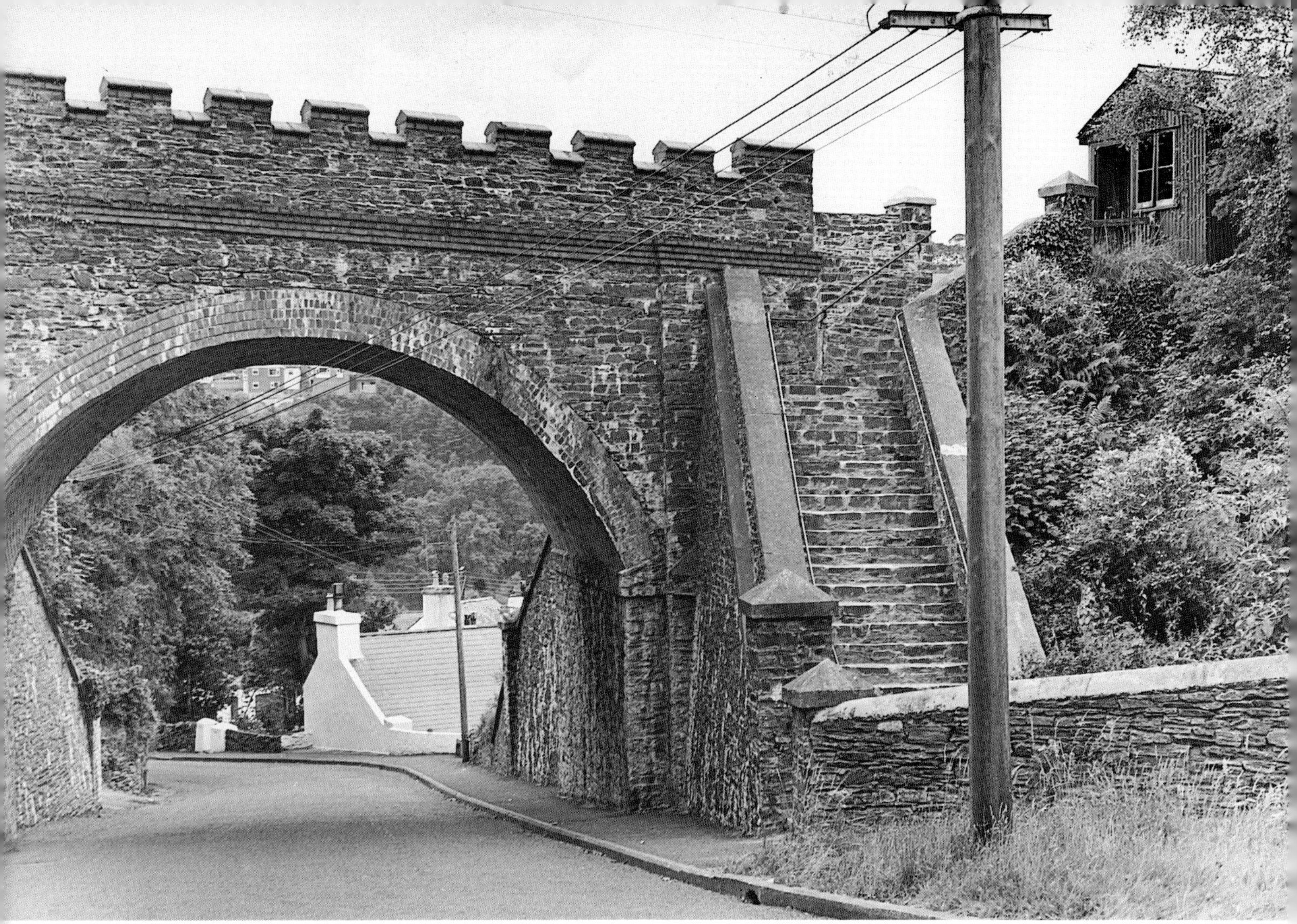

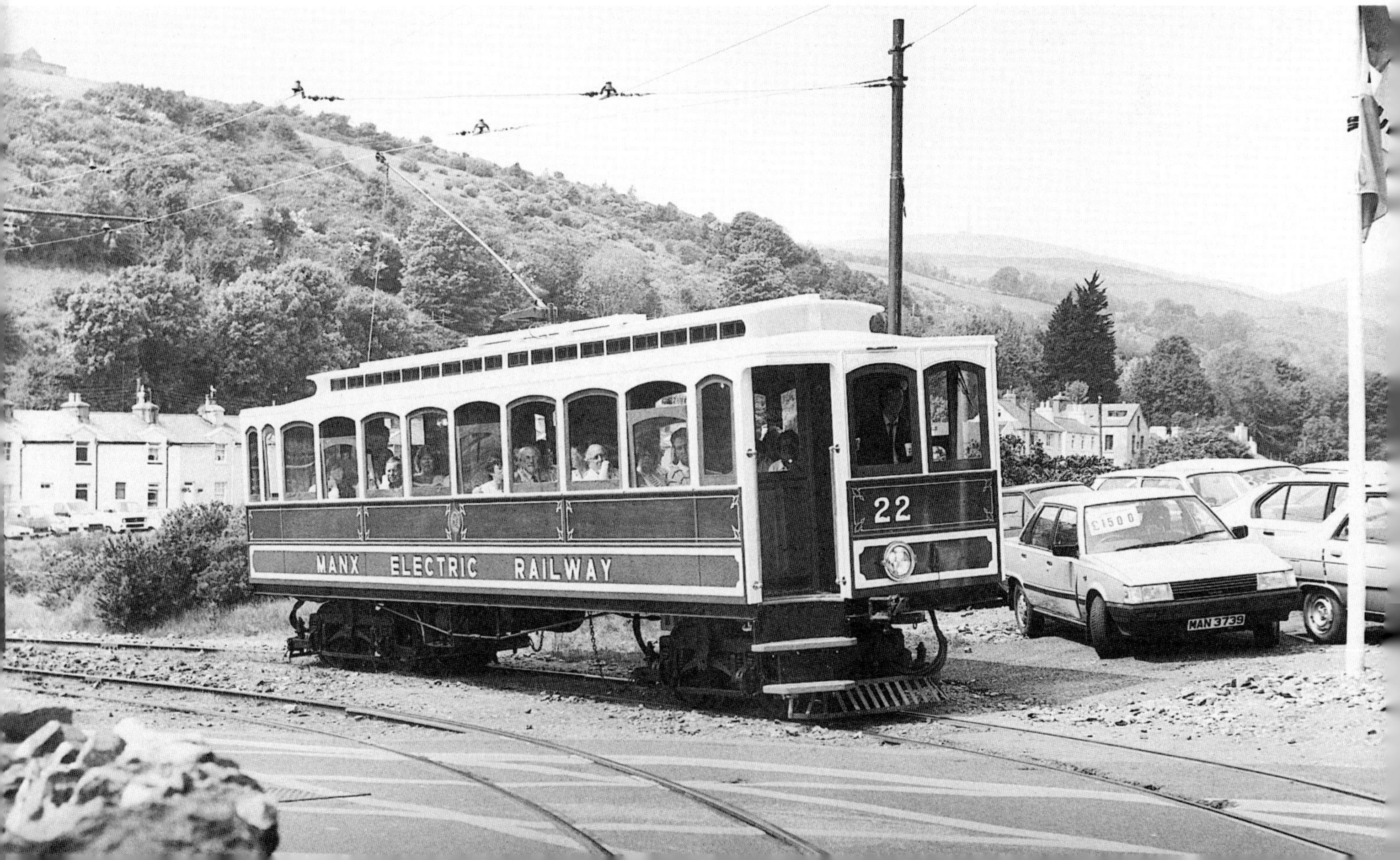
22
MANX ELECTRIC RAILWAY
MAN 3739

No. 19 with trailer 42 arrives at Laxey, having worked down from Ramsey in May 1989.

Laxey goods shed, situated at the Ramsey end of the station, was built in 1903 and is now in use as a store.

No. 6 occupies the siding alongside the goods shed at Laxey. This is used as a stabling point for cars that work out to Laxey in the morning and are not expected to return until later in the day.

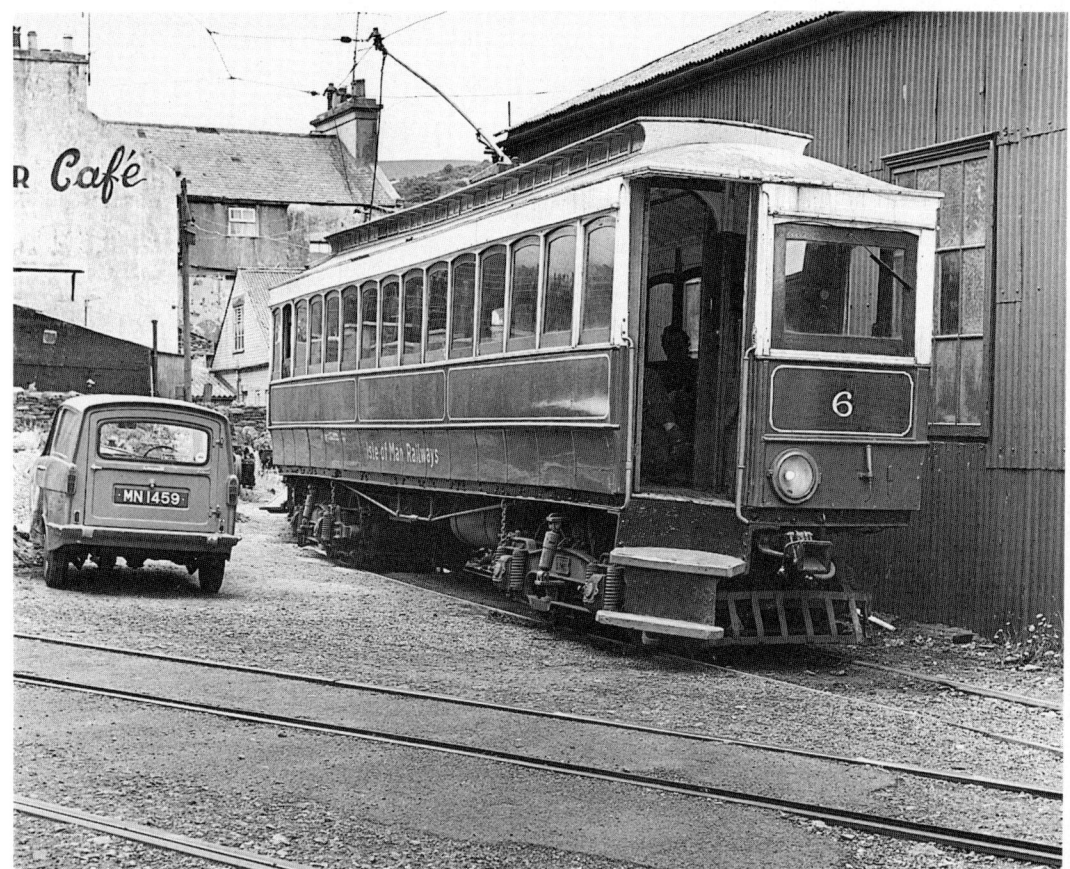

The sun does not always shine on the Isle of Man! No. 22 shelters passengers from a downpour at Laxey before setting off to Derby Castle.

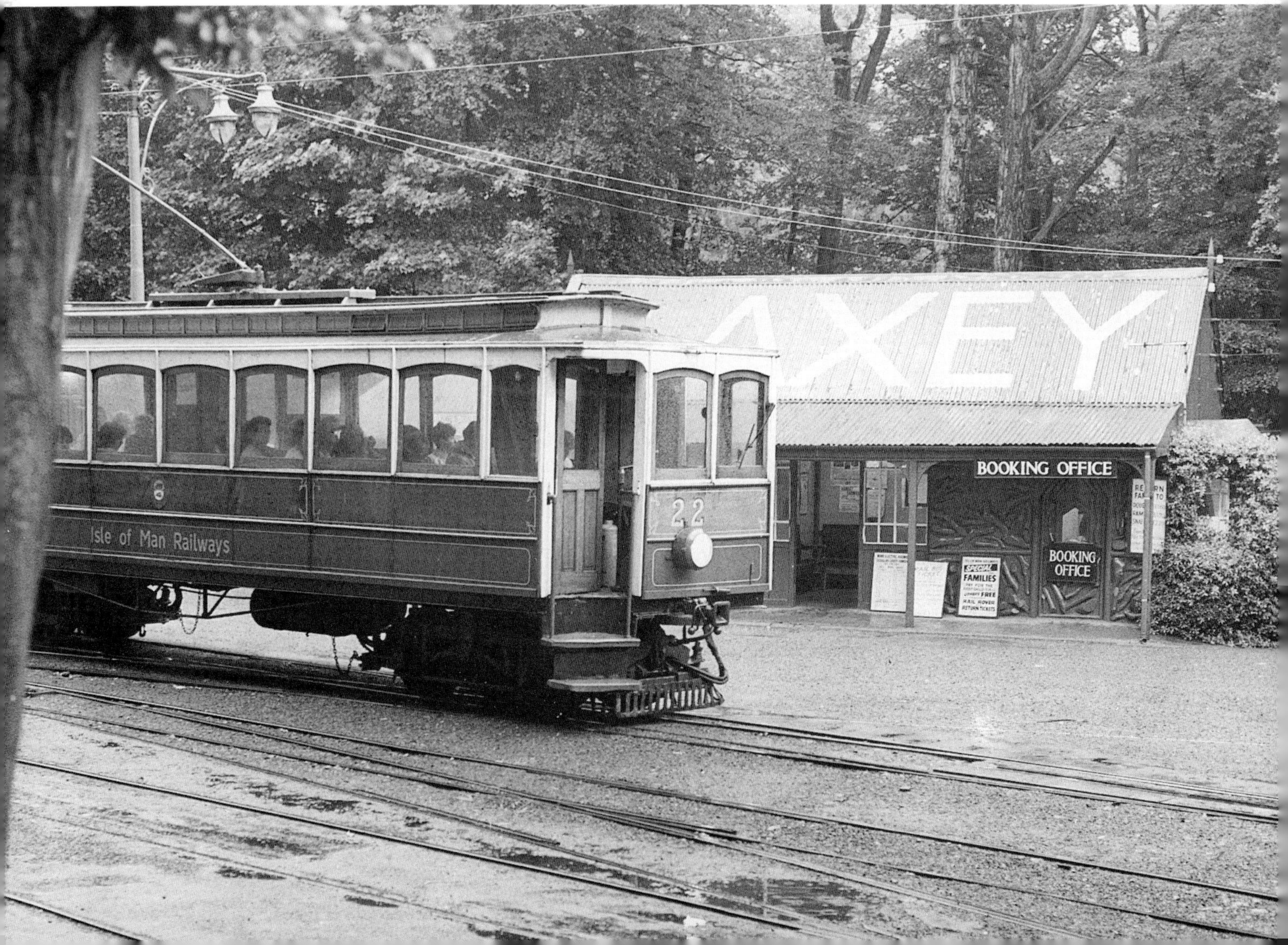

ISLE OF MAN
PASSENGER TRANSPORT
MANX ELECTRIC RAILWAY
DOUGLAS – LAXEY
1894 – 1984
TO COMMEMORATE 90 YEARS OPERATING
BETWEEN DOUGLAS AND LAXEY
28TH JULY 1984
D. F. K. DELANEY M.H.K.
CHAIRMAN

This commemorative plaque was unveiled in the Laxey waiting room to mark the 90th anniversary of through services from Douglas.

David Masters

The sun shines through the trees that surround Laxey station as open toastrack No. 32, built in 1906, arrives from Derby Castle.

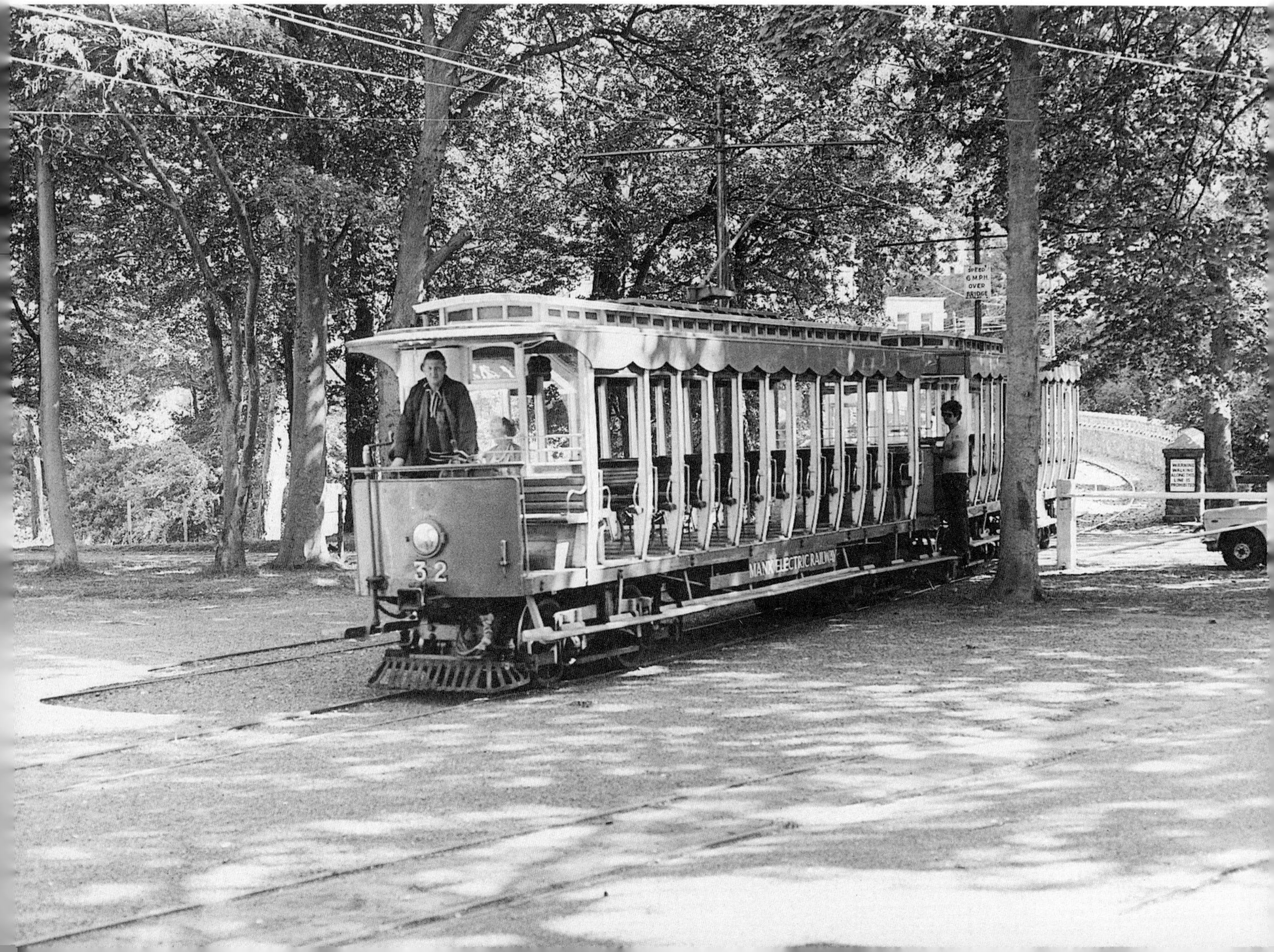

The open cars usually see service only when the line is busy and when the weather is warmer. No. 33 has just completed shunting at Laxey and is ready to depart for Derby Castle.

The southern approach to Laxey station is over Glen Roy Viaduct. No. 1 of 1893, with a well-loaded trailer, adheres to the 6 mph speed limit as it approaches the station from Derby Castle.

Shortly after leaving Laxey the line passes the sub-station built in 1934. A feeder pole is visible to the left of the building.

Left: The main road enters Laxey a little further up the glen, affording a good view of the impressive Glen Roy Viaduct. A winter saloon with trailer arrives from Derby Castle.

Right: Detail of one of the arc rectifiers in the sub-station.

Far right: An impressive looking switch panel, faced with marble, in the sub-station.

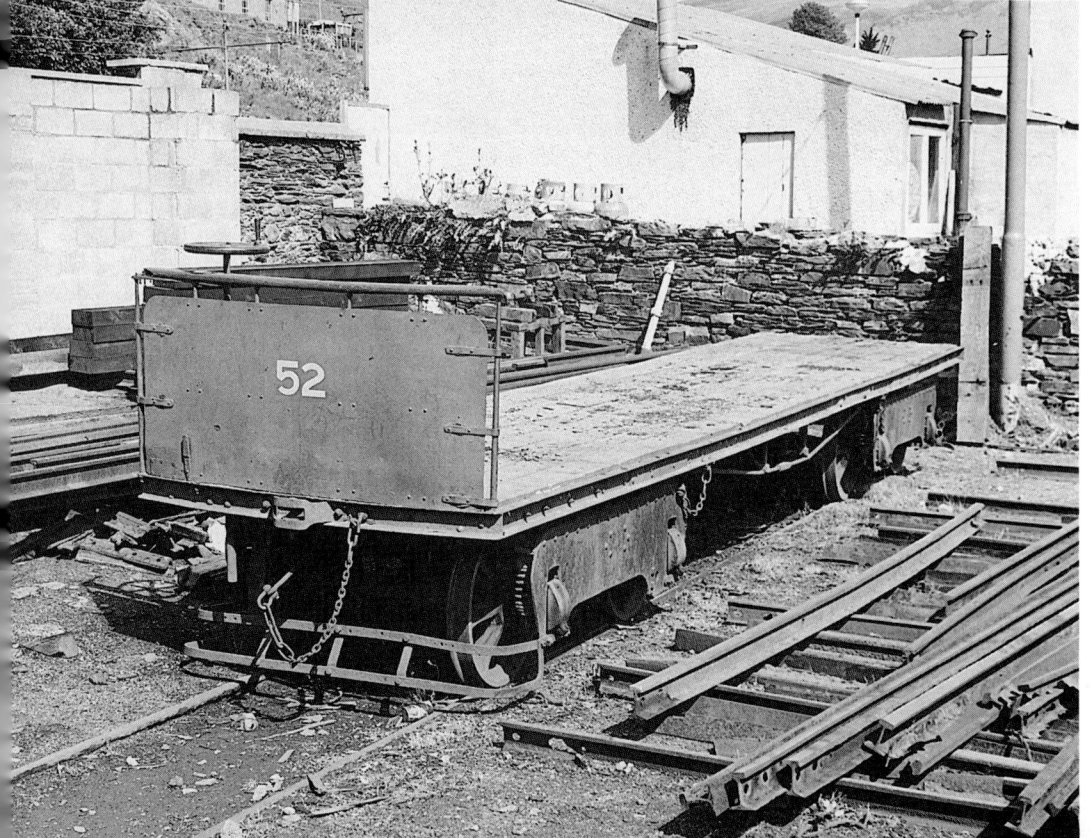

Laxey car shed is located a little nearer Derby Castle than the sub-station, the present structure replacing the shed destroyed by fire in April 1930. This picture, dated 1967, shows locomotive No. 23, trailer 50, open car 31 with a 49-54 series trailer beyond, another open car beyond the wiring trolley and car 29 to the right.

Isle of Man Railways

Trailer No. 52 of 1893 lost its bodywork in 1947 to become an engineer's flat. It is seen here occupying the shunt road at the station end of Laxey sheds

No. 9 and trailer 42 stand in the Laxey car shed headshunt in August 1981.

Laxey beach is served by South Cape station. Approaching with a service for Derby Castle is No. 33 with trailer 62.

Garwick station was once a busy stop, serving the nearby glen of the same name. The glen was sold to a private buyer, which reduced the number of passengers using the station and all the buildings were demolished by 1979, the whole area becoming very overgrown. This photograph, taken between 1903 and 1930, shows an open toastrack arriving from Derby Castle.

Author's collection

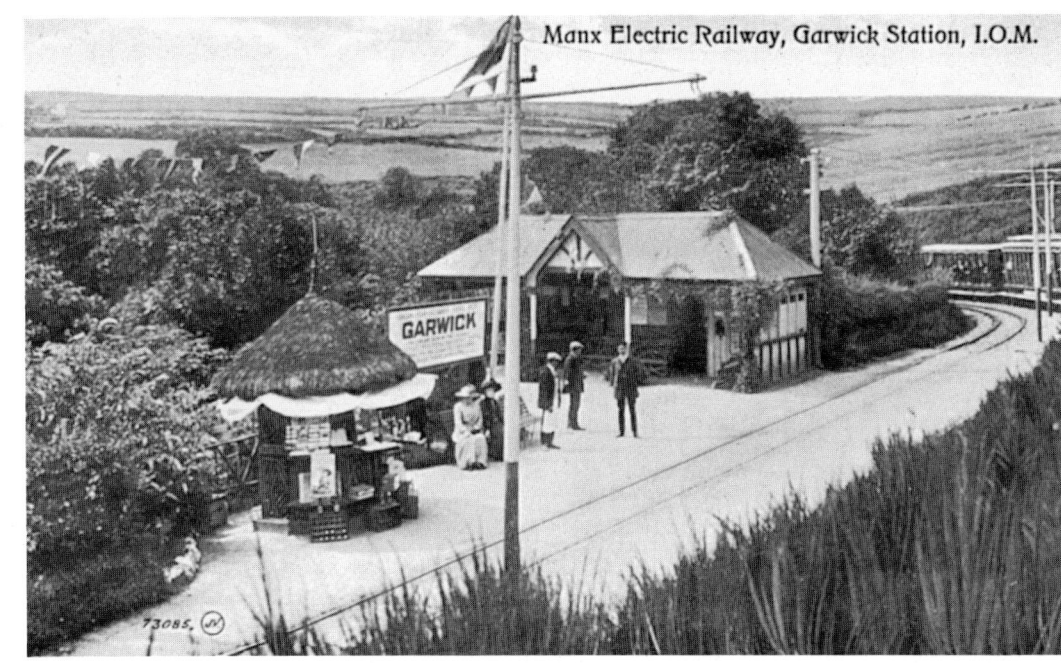

Left: The station building of Fairy Cottage can just be seen over the trees to the right of the picture as No. 9 passes with trailer 46 whilst working south towards Derby Castle.

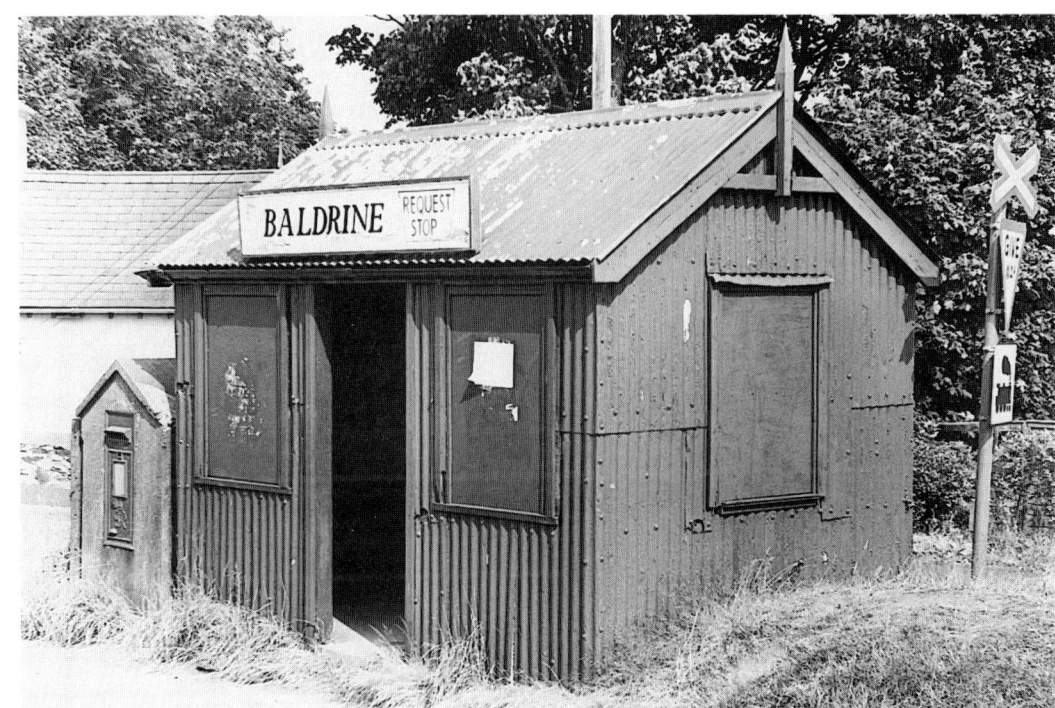

Baldrine is the next stop on the journey south. The station hut with its adjacent pillar box dates from the beginning of the century.

Approaching Baldrine station from Derby Castle is No. 5 with trailer 47. When built Nos 4 to 9 had twin front windscreens but, to improve forward visibility for the crews, these were replaced between 1966 and 1968 by the single window illustrated here.

Shortly after leaving Baldrine the line crosses the main Douglas to Laxey road. This 1951 picture shows No. 20 with trailer 48 negotiating the crossing. The traffic warning lights have since been replaced by modern level crossing type lights.

Isle of Man Railways

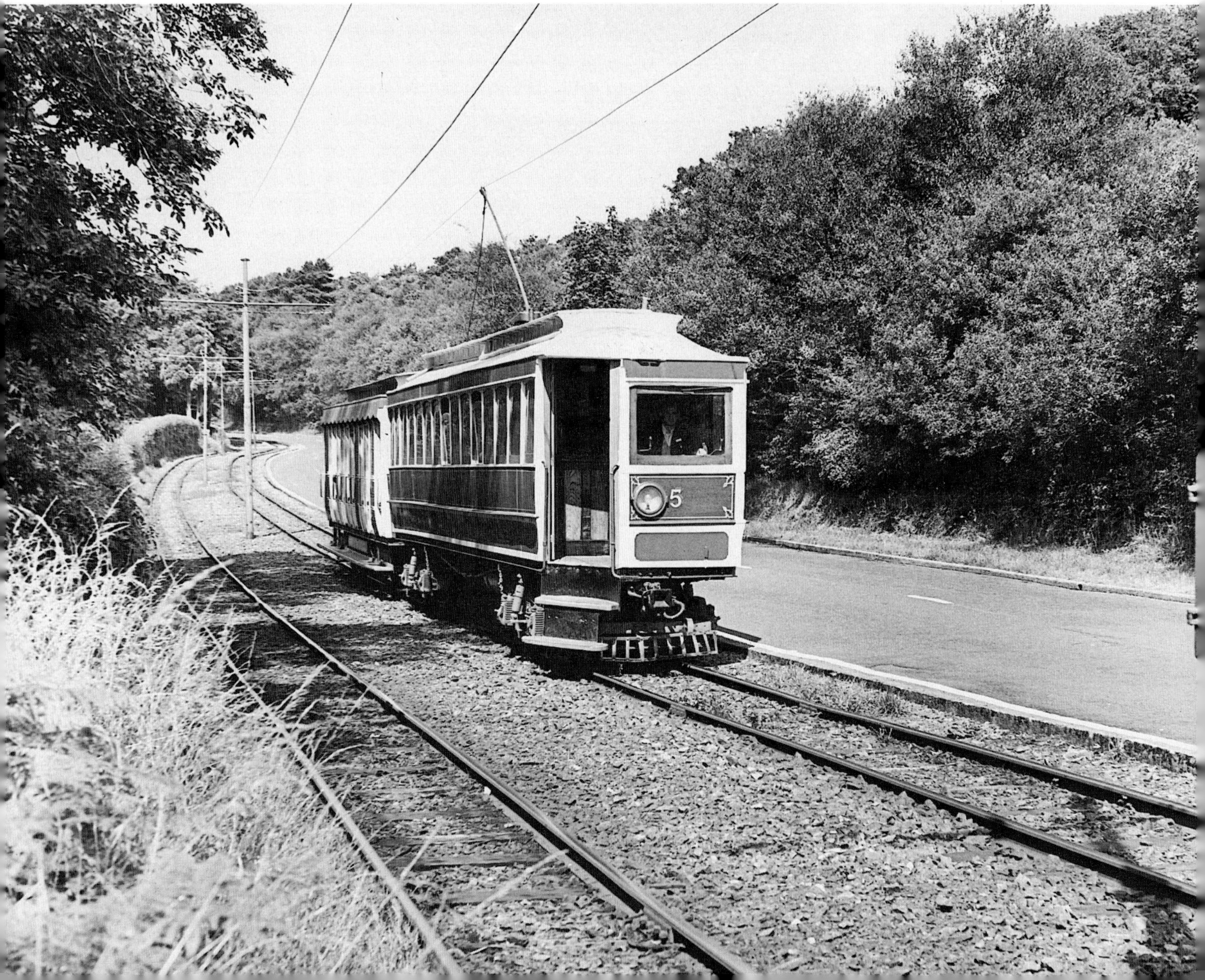

The motorman of No. 5 will soon be slowing down to negotiate the tight curves over Groudle Viaduct, as he heads towards Derby Castle with trailer 47 in tow.

The sharpest curves on the line are at Groudle where it turns through 180 degrees over the viaduct. The distance across the top of the 'U' is only about 100 yards, which includes around 20 yards of straight on the viaduct itself. No. 32 with its trailer swings into the first 90 degree curve.

A second 90 degree curve is encountered after crossing the viaduct. Here 1895 car No. 10, withdrawn in 1902, eases round the curve. Note that the car has unglazed windows. The house has recently been demolished.
Isle of Man Railways

Another 1895 car, this time No. 12, has just arrived at Groudle from Derby Castle. The elaborate glen entrance has long since disappeared but the public house is still very much open.
Author's collection

Winding their way down through Onchan village, No. 20 and trailer 45 have only about $1\frac{1}{4}$ miles to complete their journey to Derby Castle.

Beginning the steep descent into Derby Castle is No. 7 with trailer 47, bringing a full complement of passengers back into Douglas after a busy day. The White City Go Karts have since given way to a housing development.

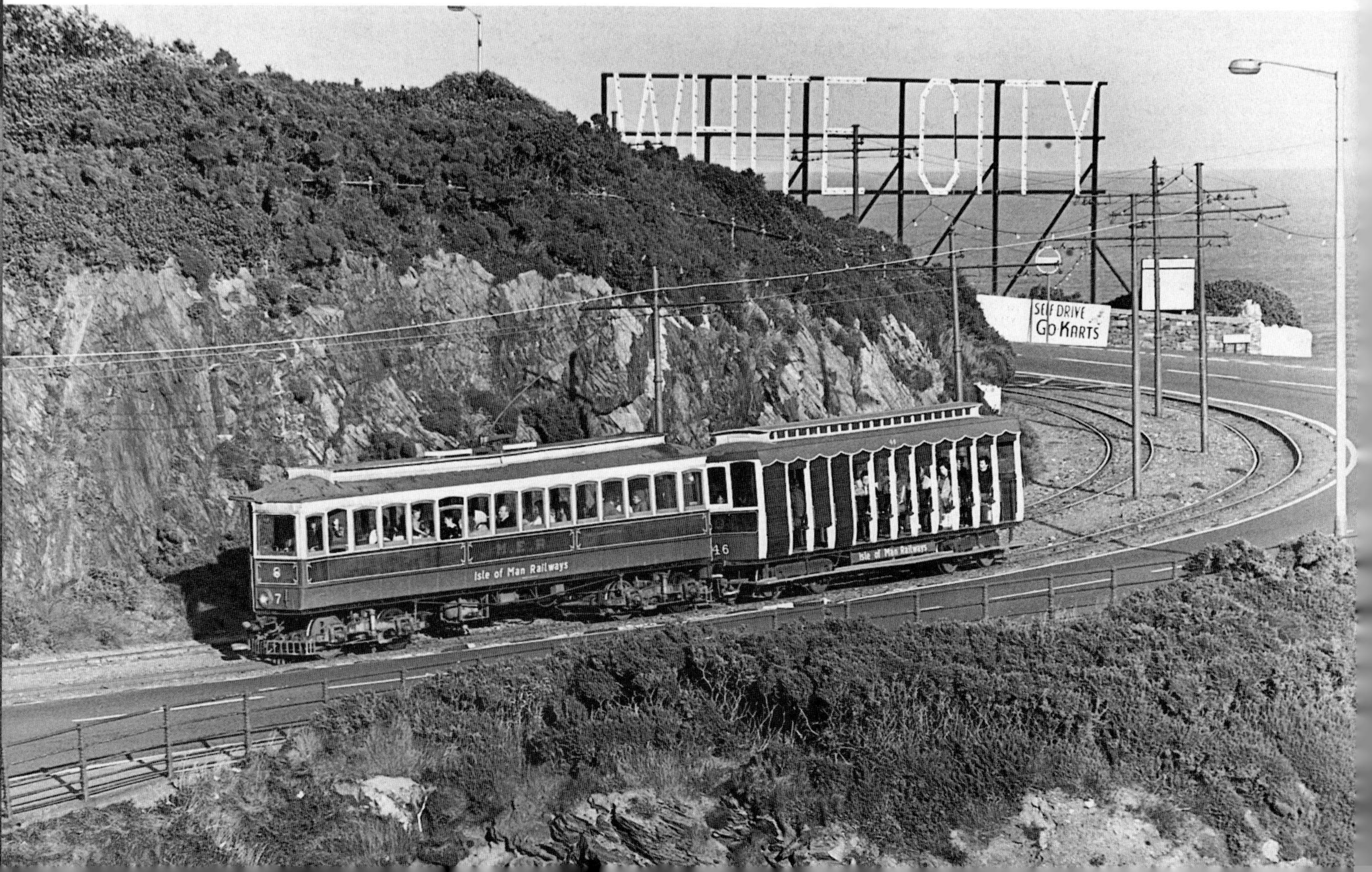

This postcard of Port Jack, posted in 1920, shows No. 32 passing the parade of shops that mark the request stop. At 1 in 23.6 this is the steepest gradient on the line.

Author's collection

Photographed from a similar position to the previous picture but some 70 years later. No. 19, with one of the line's 4-wheeled vans, heads towards Laxey with an Isle of Man Railway Society special in May 1989. The tea room is now the Port Jack Chippy.

Closed trailer No. 59 was built as a 4-wheeled unvestibuled saloon in 1895 by G. E. Milnes. It was converted to run on two bogies in 1900, which improved the quality of ride considerably and was used to carry members of the Royal Family in 1902, since when it has been known as the Royal trailer. It is seen here with No. 2 approaching Port Jack with a special working to Ramsey in May 1989.

The guard of No. 21 walks the trolley pole round the car during shunting at Derby Castle. The trailer, No. 40, will then be gravity shunted away from the photographer, before the power car backs down and makes ready for departure northwards. This scene was captured from the second of the line's overbridges, which forms an entrance to the Summerland leisure complex. Several horse trams can be seen in the right, background.

A bird's eye view of No. 21 setting off with the first trip of the day to Ramsey, passing the boundary wall of Derby Castle depot. The guard can be seen clinging to the side of the trailer whilst checking tickets.

The rustic booking office at Derby Castle, built in 1897 is now the only remaining Isle of Man Tramway & Electric Power Co. structure at the terminus.

No. 26 stands at Derby Castle coupled to 'Dreadnought' wagon No. 21 which started life as an 1895 passenger car, finally having its bodywork removed in 1926. The two Challenger cranes were added in 1977.

Photographs of the 1930 Laxey depot fire victims are rare but this one of No. 4 has been found recently. It is shown with a well-loaded but unfortunately unidentified trailer at Derby Castle. Date unknown, but clearly before 1930!

Author's collection

No. 18 with trailer 60 awaits departure from Derby Castle with an extra service to Laxey from the siding alongside the normal departure track. Terminus Building in the background houses the Isle of Man Railways' headquarters.

Darkness falls on No. 20 as it awaits departure for Groudle with an evening service during July 1985.

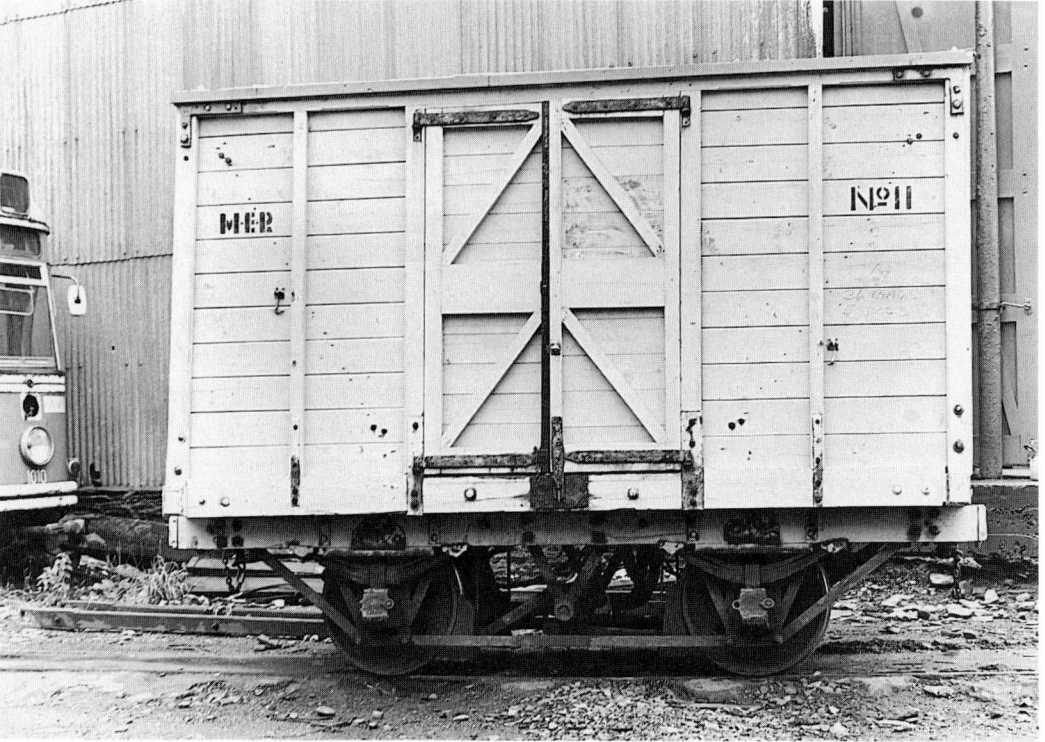

Closed trailer No. 58 is shunted in the Derby Castle depot area by one of the winter saloons. The superb condition of these trailers is partly attributable to their infrequent use.

Parked in Derby Castle depot is van No. 11 delivered in 1899. It is standing in front of one of the Aachen tramcars purchased during the 1970s to provide replacement electrical equipment for the Snaefell cars and used as a store until broken up in 1985. The tram carried the Aachen number 1010.

A partially dismantled controller column from one of the cars. The squared spindle at the top takes the main power handle, the contacts to the right being the forward, off and reverse switch to which the motorman will have a key.

An 1894 car driving position: 1, Main power control handle. 2, Forward, off and reverse switch without a key in position. 3, Air brake application handle. 4, Handbrake.

A Brill type 27cx bogie receives attention in the Derby Castle workshops. The traction motors have been removed for overhaul.

Coupling detail between a power car and trailer. The power car on the right of the picture is fitted with a Hughes Patent coupler, the trailer being connected by a specially shaped bar to take account of the varying height of trailer couplings against the power cars. The chain acts as an emergency brake for the trailer. Should the main coupling fail, the chain will be pulled tight and apply the brakes on the trailer before snapping at a specially prepared 'weak' link.

A view of Derby Castle depot from the cliff top behind shows No. 2 emerging from the sheds. Former Douglas Corporation horse trams Nos 48 and 50 stand beside No. 2. The boat, motor cycle and bus add to the variety of transport types illustrated.

The original Isle of Man Tramways & Electric Power Co. Ltd clock still ticking away on the workshop wall at Derby Castle.

Some of the wonderful old machinery still used in the upkeep of the railway.

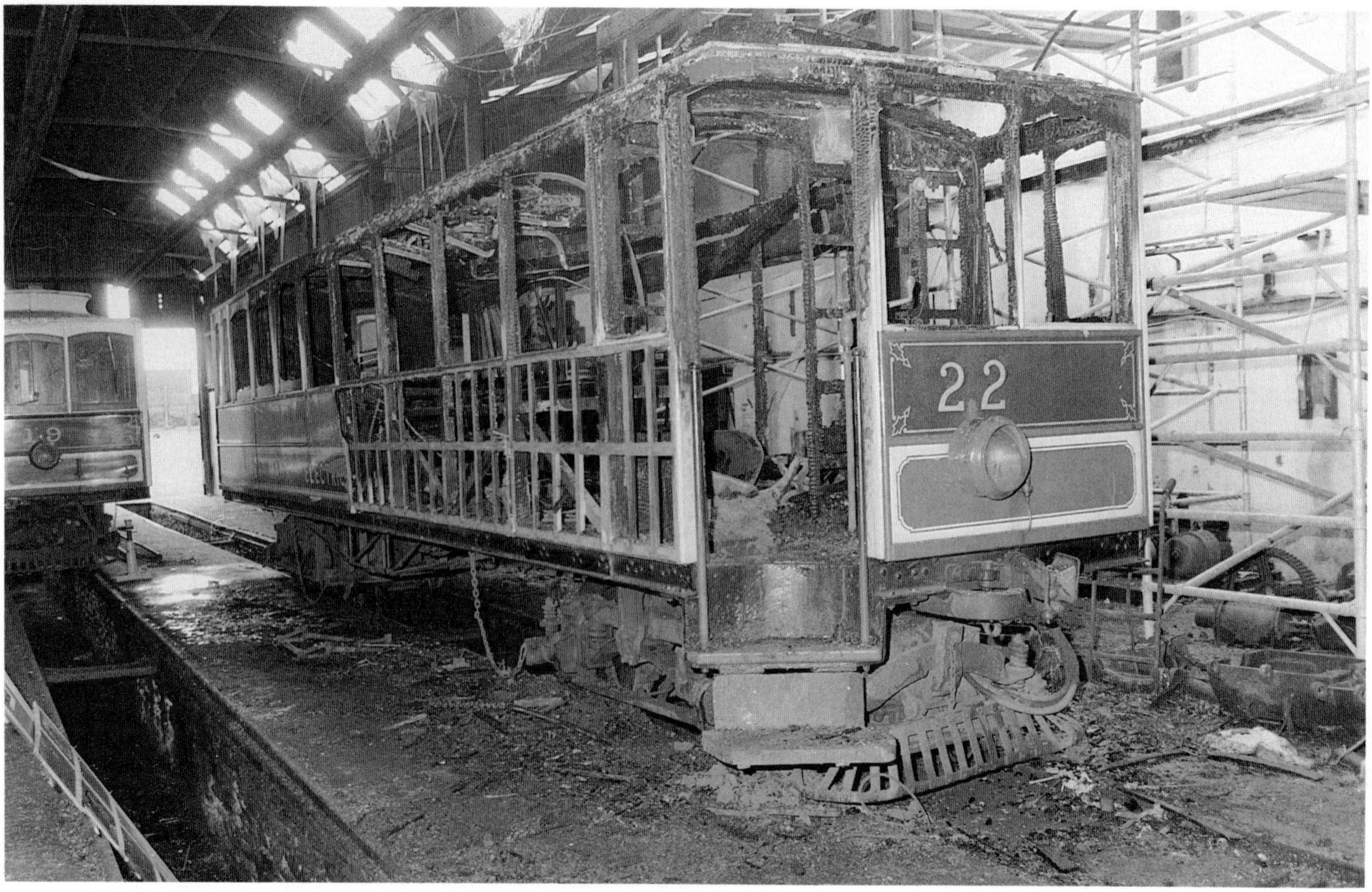

The sad sight that greeted Manx Electric depot staff when they arrived for work on the morning of 1st October 1990, following the serious fire of the previous evening. As can be seen, the fire all but destroyed No. 22; happily the car was rebuilt.

Island Photographics

The charred inside of No. 22 after the fire.

Island Photographics

A collection of trailers and a 24 to 27 series power car await their next turns of duty in the sheds. The trailer numbers from left to right are 43, 48 and 56. The uneven floor in this area is evident.

An all-important piece of equipment, the unnumbered tower wagon awaits its next call, parked with van No. 13 and the sad remains of Aachen car No. 1010 behind.

Locomotive No. 23 in its original condition at Derby Castle in 1904. It was severely damaged in a collision just north of Bulgham on 24th January 1914 and stored for over a decade.

Isle of Man Railways

No. 23 was rebuilt to this style during the winter of 1925/26 but was disused after 1944, lying in the depot for several years before being moved to store at Laxey. Rescued by the Isle of Man Railway Society, it was fully restored in 1983/84. No. 23 is seen here in the museum at Ramsey in 1979.

Dr R. Preston Hendry, who was chairman of the Isle of Man Railway Society, as well as joint author with his son of several books about the Manx railways, died in October 1991. In his memory the society decided to name No. 23 *Dr R. Preston Hendry*, the unveiling being carried out at Derby Castle on 25th May 1992.

The Snaefell Mountain Railway

The idea of a railway to the summit of the Island's highest mountain first surfaced in 1888, when Tynwald was asked to approve a plan submitted by the Douglas, Laxey & Snaefell Railway to build a steam operated line. Mr J.B. Fell, the inventor of the 'Fell Incline Railway System', was behind the project but nothing ever came of it.

The arrival in Laxey of the Manx Electric Railway in 1894 renewed interest in Mr Fell's ideas and, on 4th January 1895, The Snaefell Mountain Railway Association met in Douglas for the first time and announced its intention to build a line from Laxey to near the summit of Snaefell. The syndicate included several members of the Isle of Man Tramways & Electric Power Co., including Mr Bruce of Dumbells Bank and Mr Fell. Discussions were held about motive power and the association decided on an electric tramway. The route of the line was nearly all on Crown property, which avoided the need for an Act of Tynwald before construction could begin. In order to incorporate the 'Fell System' between the wheels the line would be constructed to 3ft 6in gauge, six inches wider than the Island's standard. Manx Northern Railway locomotive *Caledonia* was borrowed to assist with construction trains, necessitating the laying of a temporary 3ft gauge rail.

Construction of the line started in January 1895, with bad weather during February delaying the project by about a month, but by early August completion was near. Six trams were delivered from G.F. Milnes, while Mather & Platt supplied the overhead, which was erected on the entire length of the line in just ten days, the wire being 16ft above rail level.

The 4 mile 53 chain line was completed in record time, the Fell rail being laid midway between the running rails. However, during construction, it was proved elsewhere that an electric tramcar could climb a 1 in 9 gradient without assistance and so the Fell equipment was never fitted to the cars. This third rail is therefore only used for friction braking in the event of a runaway.

Various tests were carried out in mid-August and the line was officially opened from Laxey station, situated alongside the present depot, to the summit on 20th August, with the first public services on the following day. An average of 900 passengers a day were carried with the six cars operating a ten minute service.

The cars, which were unglazed, similar to the Manx Electric No. 10 to 13 series, received sliding glazed windows in 1896 and clerestory roofs during the winter of 1896/97, to stop the saloons becoming too hot!

In December 1895 the Association sold the entire line to the Isle of Man Tramways & Electric Power Co. for £72,500, £32,500 more than it had cost to build.

The original Summit Hotel, located a little to the south of the terminus, was extended in 1896 and a new hotel was built at Halfway, now known as Bungalow.

The Isle of Man Tramways & Electric Power Co. considered the distance between their coastal line station and the Snaefell station in Laxey to be too great. The Snaefell terminus was therefore moved nearer in 1897, and in 1898 a further move brought it into the same station as the coastal line, as is still the case.

The Snaefell line was sold to the Manx Electric Railway Co. in 1902, following the collapse of Dumbells Bank. As with the coastal line, much re-equipping was needed and changes were made to the generation and distribution of electricity.

The only serious incident to occur on the line was on 14th September 1905, when three cars were descending in convoy. The first stalled, the second stopped without difficulty but the third failed to stop, colliding with the second and pushing it forward into the first. All three cars were damaged and a number of passengers injured.

The line continued to be a considerable success and the hotel at the summit proved inadequate to cope with the vast numbers of passengers. In order to overcome this, an elaborate new hotel was built alongside the railway terminus, opening in 1906.

A note in Isle of Man Tramways & Electric Power Co. documents of 1897 considered the possibility of a line from Bungalow to Tholt-y-Will but nothing came of this. In 1907 however, following the construction of a hotel and refreshment room at Tholt-y-Will, the Manx Electric Railway Co. started a motor charabanc service over this

The original Snaefell station shortly after the arrival of a car from Laxey. The actual summit is above the car to the right of the picture. The station was replaced by a hotel in 1906.
Manx Museum and National Trust

route. This and mountain railway services ceased at the outbreak of World War I.

Following the end of hostilities and the clearing of arrears of maintenance, the line reopened on 10th June 1919. The Tholt-y-Will service also restarted but with different vehicles, as the charabancs had been sold. Two Ford Model Ts took over in 1926 and these, in turn, gave way to Bedford coaches in 1939.

All rail and road services were stopped on 20th September 1939, soon after the outbreak of World War II, but the line saw some traffic after this, carrying peat down to Laxey from the Bungalow area to assist with fuel shortages. This traffic was handled by freight car No. 7, nicknamed "Maria", with bogies borrowed from passenger car No. 5.

The post-war boom in the tourist industry on the Island provided the line with plenty of passengers and frequent services were operated. Visitor numbers began to dwindle in the early 1950s but a decision by the Air Ministry (later the Civil Aviation Authority) to construct a radar station at the summit provided work for the line during the winter of 1950/51. The radar station presented a problem in that it had become customary to remove the overhead wire from the upper section in winter to avoid damage. As the Ministry would need the line all year, it purchased a Wickham railcar in 1951 and

The new hotel was an impressive building, seen here with one of the mountain railway cars displaying the Snaefell Mountain Tramway lettering, sometime before 1921.
Author's collection

Manx Electric Railway Hotel, Summit of Snaefell. I.O.M. (2034 feet above Sea Level.)

From the summit of Snaefell you can see five nations: the Republic of Ireland, Northern Ireland, Scotland, England and Wales, but only when it is clear! With the mist swirling round, No. 1 awaits departure for Laxey with a full complement of passengers.

SNAEFELL MOUNTAIN RAILWAY

The unusual pointwork at the summit station in which the centre rail swings right across. For safety reasons the cars run on the right and so the point is set for the arrival of a car from Laxey.

a second in 1957. These are housed in a small shed on the Snaefell depot site.

Passenger numbers continued to fall and in 1953 the coach service to Tholt-y-Will was withdrawn. At the end of the 1955 season the Manx Electric Railway Co. advised the government that it would be unable to operate the railway after the end of the following year. However, a nationalisation agreement was reached and the Manx Electric Railway Board took control of the line on 1st June 1957.

The Tholt-y-Will coach service was reinstated but with little patronage, was extended to Sulby the following year and soon abandoned altogether. The Summit Hotel was redecorated in 1958, whilst the Bungalow hotel was closed and demolished. The green and cream livery was applied to cars 2 and 4, No. 4 being the last car of either the mountain or coastal lines so treated.

The mid 1960s saw the entire line showing signs of age and corrosion, the Fell rail being of particular concern. Whilst the track was receiving attention, the cars themselves were becoming in desperate need of mechanical and electrical overhaul and, between 1958 and 1975, several possible solutions were considered and rejected for various reasons.

Disaster struck on 16th August 1970 when, shortly after arrival at the summit on a particularly wild day, car No. 5 caught fire and was almost completely destroyed. It was later discovered that a high tension cable had rubbed

On a clear day the views from the Snaefell line are superb. Here, No. 3 ascends the mountain past the Sulby reservoir and the rolling hills beyond.

The Snaefell line has its own depot and light maintenance facilities at Laxey. Seen here in the depths of the shed is No. 6 which has a day off while cars 1 to 5 ferry passengers up and down the mountain. The depot has just two tracks, each long enough to hold three cars.

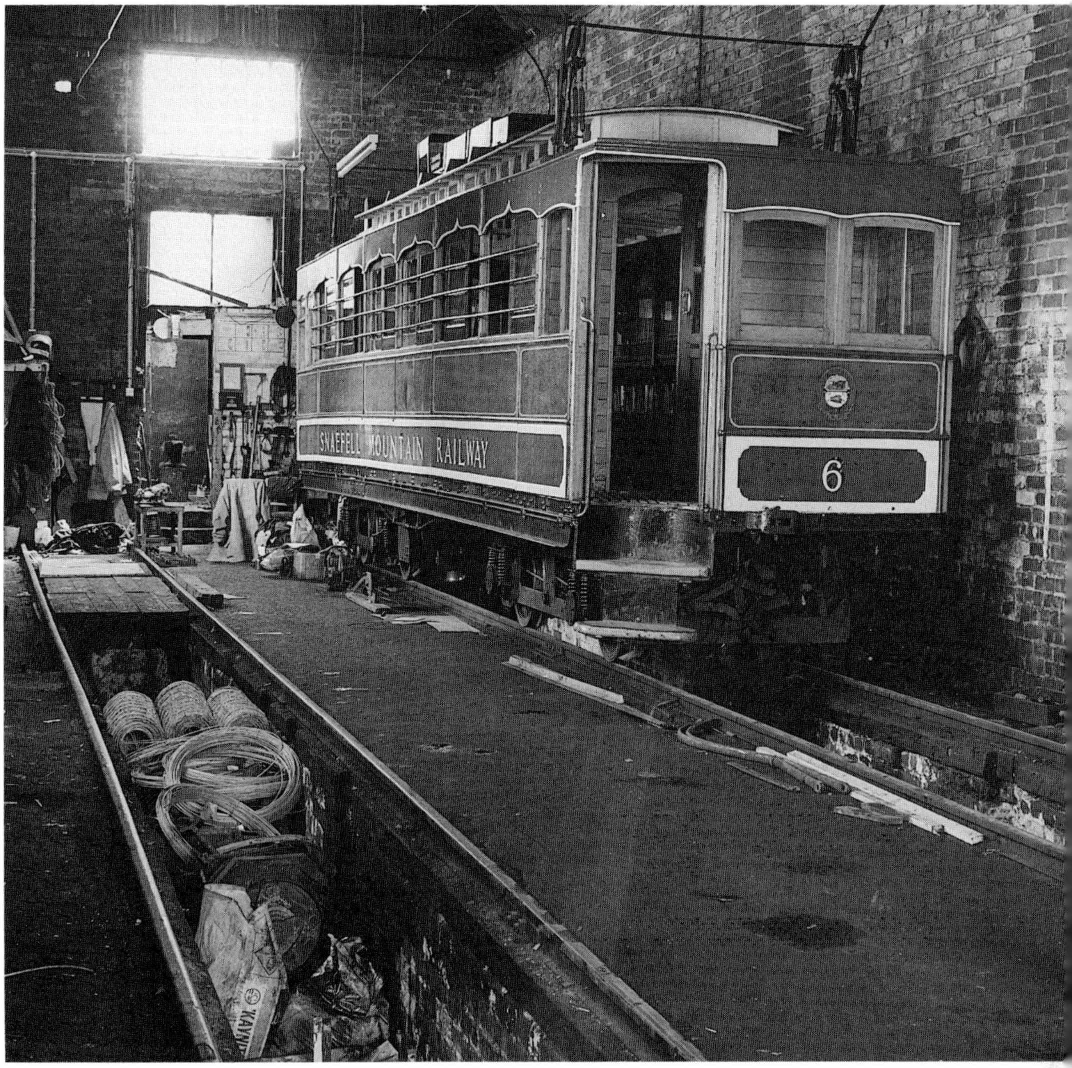

Left: Photographed from a descending car, No. 6 starts its 1 in 12 ascent of the mountain shortly after leaving Laxey.

A general view of the Snaefell depot, with the line's wiring trolley parked alongside the shed. The shed on the left houses the Civil Aviation Authority Wickham railcars, used in winter to gain access to the radar station on the mountain summit.

One of the Wickham railcars parked alongside the Snaefell car shed.

Detail of the Fell brake shoes used on the Snaefell cars. One of the 61hp motors is also visible.

Before crossing the road and entering the station at Laxey, the Snaefell line becomes single track. Here, No. 6 has left the station and is working 'wrong line' back to the depot.

Snaefell car No. 5 stands at Laxey awaiting departure to the summit. The bus style windows fitted during the rebuilding of this car are clearly visible. There is a marked similarity between the Snaefell cars and the Manx Electric 10 to 13 series, two of which were stored for a long while following their withdrawal in 1902, with the possible intention of converting them to additional Snaefell cars.

To assist with the freight traffic generated by the rebuilding of the Summit Hotel following the 1982 fire, two of the discarded 1895 bogies were fitted with small wagon bodies. One of these, No. 1, is seen standing in Laxey station, with a large water tank aboard.

The Snaefell wiring trolley at Laxey. The line has to have its own trolley because its gauge is different to the Manx Electric Railway.

Arriving at Laxey on a clear sunny morning it is not unusual to find three or four Snaefell cars lined up ready for their day's work. One of these has a small wagon, carrying supplies for the Summit Hotel, which it will propel up the mountain because the wagon has no brakes!

Isle of Man Railways and Tramways in Colour

Manx Electric car No. 19 skirts the coastline at Bulgham Cliffs, whilst working from Ramsey to Douglas during May 1991.

The replacement Car 22 stands in the tropical setting of Laxey station on Sunday 24th May 1992, with a special working to Ramsey.

On 5th December 1991 Steam Railway locomotive No. 4 *Loch* visited the Manx Electric for tests, prior to the centenary celebrations to be held in 1993 when a steam locomotive will work from Laxey to Dhoon on several dates. This picture, taken in the rain, shows Manx Electric No. 1, Steam Railway No. 4 and Snaefell Mountain Railway No. 6, side by side at Laxey station.

Albert Lowe

Following the destruction of No. 22 by fire, the decision was taken to rebuild the car. The 'new' 22 takes shape in the depot at Derby Castle during May 1991.

Manx Electric Car No. 2, dating from 1893, stands at Derby Castle with trailer No. 13. Note that the trolley pole is facing the wrong way and must be reversed before the car departs.

As it ascends towards Bungalow and the summit, Snaefell car No. 3 is framed in the front windscreen of a Laxey bound car.

Carol Edwards

The superb Manx coastline forms a backdrop for the Groudle Glen Railway's *Sea Lion*, running round its train at Sea Lion Rocks station on 24th May 1992, the day after services were restored to this terminus.

Sea Lion winds its way out of the woods and approaches Lime Kiln Halt with a Headland bound train in May 1988.

Groudle Glen Railway diesel locomotive No. 1 *Dolphin* stands outside the railway's depot at Lhen Coan. One of the Fauld wagons is visible behind the locomotive.

No. 4 *Loch* of the Isle of Man Steam Railway prepares to leave Douglas with the 10.10 to Port Erin on a bright sunny morning in May 1988.

The steam railway's Schoema 4-wheel diesel locomotive stands outside the Douglas workshops in August 1992 shortly after arrival from Germany. The vast German couplings have been removed, as has the builder's plate from this side.

Island Photographics

Former County Donegal diesel railcars, numbered 19 and 20 by the Isle of Man Railways, approach Castletown on a special working in May 1986.

No. 12 *Hutchinson* arrives at Castletown with a Port Erin to Douglas train in May 1988. The driver is handing the single line token to the stationmaster.

The Groudle Glen Railway

During the late 19th century Groudle Glen was being developed as a tourist attraction. A zoo was built at the mouth of the glen along with a refreshment room, and a further refreshment room was constructed just below what is now Headland station. The Groudle Glen Railway was conceived to transport visitors from Lhen Coan station, situated near the entrance to the glen, out to the refreshment rooms and zoo.

The 2ft gauge railway was constructed by local labour during 1895 and the early part of 1896. Spring 1896 saw the arrival of the railway's first locomotive, *Sea Lion* and three 4-wheeled coaches. The railway opened to the public on 23rd May 1896, with 10,000 passengers being carried in the first two months and over 100,000 in the first three years, with up to 40 return trips being worked each day.

During the winter of 1904/5 a new longer shed was constructed at Lhen Coan station and a run-round loop installed at Headland, in readiness for the arrival in May 1905 of the second steam locomotive, *Polar Bear* and four further coaches.

The outbreak of World War I caused the railway to close but it soon reopened at the end of hostilities and visitors flooded back to the glen. Sadly a steep rise in coal prices caused the two steam locomotives to be stored and replaced by battery electric machines. These proved troublesome, one of them finishing up at the bottom of the glen following an accident. After only six years the electrics were withdrawn and the two steam locomotives overhauled by their builders and returned to service.

World War II caused the railway to close once again, not reopening until 1950, and then only to the Headland because of a serious rockfall between there and Sea Lion Rocks. *Polar Bear* was the only servicable locomotive, with six coaches available. A small petrol locomotive was tried out on the line as a spare but soon disappeared as it was too weak for the gradients and uncontrollable on downward sections. The 1950s saw a decline in the number of visitors and an irregular service operated.

A change of ownership of the glen in 1960 resulted in a revival, with *Polar Bear* and the six coaches operating with regular services until the end of the 1962 season. During the 1962/3 winter a good deal of repair work was carried out and attempts were made to purchase a second-hand locomotive but none could be found. The 1963 season came and went but the railway was unable to operate any trains, due to the failure of *Polar Bear*. The 'old' railway never reopened.

In 1965 the Groudle Glen Railway Preservation Society was formed. At first things looked promising but in 1967 the owners of the glen ordered the removal of all railway equipment, with the result that, by June 1968, the locomotives and rolling stock were dispersed all over the British Isles. *Sea Lion* and a few coaches managed to stay on the Island, being exhibited at the Steam Centre in Kirk Michael and these items were left behind when, in the late 1970s, the Steam Centre closed down and moved to the newly established Midlands Steam Centre at Loughborough. This little line had surely gone for good.

The Sea Lion Locomotive Association was formed in

Sea Lion stands at Sea Lion Rocks station on Sunday 24th May 1992, the day after services were officially restored to the terminus after an absence of 53 years.

The original buffer stops at Sea Lion Rocks station on the Groudle Glen Railway photographed in 1979. The old rails can just be seen in the foreground.

1981, eventually moving their locomotive to Loughborough in October 1981, where it was restored externally.

In January 1982 the Isle of Man Steam Railway Supporters Association embarked on what must be one of the most impressive narrow gauge railway restorations ever. Structural surveys were carried out and it was agreed that to reinstate the railway was possible, providing the original formation was kept. The new owners of the glen and the Manx government gave full support, and planning permission was received in May 1982.

The Supporters Association members needed no further encouragement and restoration work started, clearing the trackbed of fallen trees, gorse and bracken being the first job. A borrowed mechanical digger meant that what would have been months of clearing work was completed in a matter of weeks, while a smaller digger was used to dig trenches for new drains to be installed.

A few 4-wheeled wagons formerly used at RAF Fauld, Staffordshire were purchased and arrived on the Island in October 1982. Spring 1983 saw the purchase of the entire Dodington Narrow Gauge Railway, situated near Chipping Sodbury, Avon, which had gone into voluntary liquidation. This provided two diesel locomotives, two coaches plus a spare frame, two points and 1,080 yards of track.

Sea Lion was also the subject of negotiations with her owners and it was agreed that the Association could complete the restoration, on which little progress had been made at Loughborough, and operate the locomo-

tive for as long as they continued to run the railway. This locomotive returned to the Island in March 1983. Meanwhile, in the glen, 150 tons of ballast had been delivered and tracklaying was completed from Headland to Lime Kiln Halt.

Towards the end of 1983, the volunteers could wait no longer. They set up a Santa's grotto in the engine shed at Headland and on 18th December 550 people rode between Lime Kiln Halt and the grotto, the first passengers to travel on the line for 21 years. Two of the Fauld wagons were used, hauled by one of the diesel locomotives, specially named *Rudolph* for the day. Boxing Day saw more Santa specials and a further 200 visitors.

Tracklaying had reached half way to Lhen Coan by mid 1984 and, as the Apprentice Training Centre at

Ravenglass & Eskdale Railway volunteers tracklaying at Sea Lion Rocks station during a working visit in February 1992. The original alignment can be seen on the left.

Sid Edwards

The same scene photographed earlier in May 1991. The massive amount of earthworks that had to be carried out is clearly evident. The new pointwork patiently awaits the first train.

BNFL Sellafield had offered to carry out all remaining restoration work on *Sea Lion*, the locomotive left the Island once again in October 1984 bound for Cumbria. Autumn that year saw considerable efforts made to complete the tracklaying as far as Lhen Coan station, which was duly achieved in readiness for a repeat of the Santa trains. The second diesel locomotive was delivered to the line, along with the first coach, built on one of the former Dodington underframes. The two diesels were named *Dolphin* and *Walrus* and the yuletide running was another outstanding success.

Work continued through 1985, a wet summer delaying further deliverics of ballast, while Tynwald passed the Groudle Glen Railway Order which meant that the railway had to be inspected before any more passengers could be carried. The inspector called in November and issued a pass certificate, opening the way for 1,300 people to visit the line over Christmas.

February 1986 saw the steelwork for the new shed arrive in the glen and, by the end of March, the 20ft by 50ft structure was in place. A second coach was delivered and the temporary shed moved underneath the steelwork of the new one.

On 25th May 1986, to the delight of the Isle of Man Steam Railway Supporters Association volunteers and to the sound of Rushen Silver Band, *Dolphin* was driven through the white tape by Mrs Carolyn Rawson, daughter of the late Dennis Jeavons who had conceived and developed the Groudle Glen holiday village. After 24 years the Groudle Glen Railway was now officially open again, probably in better condition than it had ever been before. Trains ran every Sunday for the rest of the year, the Santa trains attracting 1,600 people.

Early 1987 was spent completing the new running shed, ready for the season, the highlight of which was the return of *Sea Lion* to the railway on 8th September.

Steam was raised for the first time in 25 years on 21st September and an official handing over ceremony took place on 3rd October. The now annual Santa specials were steam hauled for the first time.

Restoration work continued between running trains with, in 1988, the completion of run-round facilities at Lhen Coan and thoughts turning to the reinstatement of the line to Sea Lion Rocks. Cosmetic work included a start on the building of a replica station canopy at Lhen Coan, which is still under way at the time of writing.

During late 1990, after much negotiation and the purchase of a strip of land from a local farmer, work started on the extension of the line to its original terminus at Sea Lion Rocks. A mechanical digger was called in once again to shift tons of soil, moving the railway ledge further inland. Tracklaying commenced in early 1991, continuing throughout the year and on into 1992. New fences were erected around the cliff edges at Sea Lion Rocks to prevent any accidents, the finishing touches being made to the trackwork during March. Diesel locomotive No. 1 *Dolphin* operated a test run over the new track on 29th March, following which permission was given by the Railway Inspector for public services to commence.

Test runs in early April brought the return of *Sea Lion* to the old terminus for the first time in 53 years, with public services re-commencing on Easter Sunday 19th April. Saturday 23rd May 1992 saw the official re-opening of the extension by Mr James Cain, Speaker of the

Throughout its length the Sea Lion Rocks extension from Headland runs approximately six to eight feet further inland than the original line. The new soil face created by the earthworks has yet to produce vegetation as *Sea Lion* passes with a Lhen Coan bound train.

Sea Lion at Headland having just arrived from Lhen Coan during May 1988. The excellent job done by the Sellafield apprentices is clearly evident in this picture.

Two of the ex-RAF Fauld wagons have been converted to form a well-equipped works train and are seen here leaving Headland for Lhen Coan, behind No. 1 *Dolphin* in May 1991.

On 25th May 1986, the first day of public service, No. 1 *Dolphin* works a train round the Headland curve towards Lhen Coan. The first coach has since received a roof to match the second. The driver of No. 2 *Walrus* can just be seen at the rear of the train.

Seen approaching Lime Kiln Halt is one of the short-lived battery electric locomotives hauling four coaches. Although undated this picture must have been taken between 1921 and 1927.

Isle of Man Railways

The section of line from Lime Kiln Halt to Lhen Coan is all in woodland. Entering the woods shortly after leaving the halt is *Sea Lion* with the railway's two coaches and a good number of passengers.

Sea Lion trundles through the woods with a train for Headland during May 1991. The vast trees rather dwarf the 2ft gauge locomotive and its train.

The steelwork for the sheds at Lhen Coan has been completed, the temporary wooden shed having been moved from over the running line on the left to under the steelwork. On completion of the new shed, the wooden one was dismantled. The pointwork for the left hand shed road can just be seen, the right hand road is already laid. 25th May 1986.

The scene at Lhen Coan in August 1979, with some of the canopy uprights still visible. When the area was cleared for reconstruction a number of small parts from the original line were discovered in the undergrowth.

The same scene during 1991 with the framework for the replica station canopy in situ. Note that the coaches have no doors on the landward side.

Hundreds of people gathered in the glen on 25th May 1986 to witness the official opening of the line. Here, No. 1 *Dolphin* is driven through the white tape by Mrs Carolyn Rawson, daughter of the late Dennis Jeavons, founder of the holiday village.

Close-up of the first covered coach at Lhen Coan station. Each coach is fitted with a handbrake at one end.

Lhen Coan station in 1900, with *Sea Lion* awaiting departure for Sea Lion Rocks. The lack of trees is evident, compared with recent photographs of the same scene.

Manx Museum and National Trust

As mentioned earlier in this section, the other Groudle locomotive has also been preserved. This picture shows *Polar Bear* in operating condition at Amberley Chalk Pits Museum in West Sussex in August 1983.

David Masters

One of the preserved Groudle coaches fully restored at Amberley Chalk Pits Museum.

The Douglas Horse Tramway

The Douglas Horse Tramway operates today on a 1.6 mile 3ft gauge track laid down the middle of the Promenade from Derby Castle at the northern end to the Sea Terminal at the southern end.

The Tramway was first conceived in 1875 by Thomas Lightfoot, a retired civil engineer from Sheffield who was involved in the construction of the original Woodhead Tunnel, and later that year he lodged a proposal at the Rolls Office in Douglas. Spring 1876 saw an Act of Tynwald and the granting of Royal Assent for a tramway from Victoria Pier, part of the present sea terminal, to Summer Hill. The Act specified that only animal power could be used.

A single track line with passing loops was constructed with 35 lb/yd rail, the centre walkway for the horses being laid with small stones and tar. The Public Highway Surveyor inspected the line and it opened, without ceremony, between Summer Hill and the Iron Pier, situated opposite the bottom of Broadway, on 7th August 1876.

The Starbuck Car & Wagon Co. Ltd, later G.F. Milnes & Co., delivered two double-deck tramcars, only one of which was actually operated on opening day, hauled by two horses.

January 1877 marked the opening of the second stretch of line, linking the Iron Pier with Victoria Pier. After experiments with just one horse, public opinion obliged the operators to continue to use two horses on the double-deck tramcars. Later that year the stable building that is still in use today was purchased to house the expanding stud of horses. In 1882 the Horse Tramway was sold to Isle of Man Tramway Ltd, which added two more passing loops in 1883, in order to increase service intervals to every 20 minutes. This was still insufficient and in 1884 a further passing loop was added and the tram fleet increased to eight.

Over 350,000 passengers enjoyed a ride on the line in 1885 and two more single-deck trams were added. Summer Hill terminus was rebuilt and renamed Derby Castle

The peaceful scene that greets horse tram passengers at Derby Castle. Passengers and horse await departure with No. 33 to the Sea Terminal. Tram 46, seen to the left, was withdrawn from service in 1987 and has since been sold for preservation at Birkenhead.

in 1886, permission being granted in the same year to double the track from Falcon Cliff to Derby Castle. Seventeen tramcars were now in the fleet and they all took part in the ceremony to mark the opening of the double-track section in 1887. In 1888 over half a million passengers were carried and 79,278 tram miles covered.

Twenty-six tramcars were in service by 1891 and all but a short stretch of line had been doubled. Thomas Lightfoot died on 10th January 1893, just eight months before The Douglas & Laxey Coast Electric Tramway Co. opened the first stage of what is now the Manx Electric Railway, from Derby Castle to Groudle. The electric tramway terminus at Derby Castle, just fifteen feet away from the horse tram terminus, brought even more passengers to the Horse Tramway.

The Isle of Man Tramways & Electric Power Co. purchased the Horse Tramway for £38,000 in April 1894 and in 1895 work commenced on the new tram depot at Derby Castle, which is still in use today. The offices on top were added later and now house the Department of Tourism and Transport. The cast iron awning over the Horse Tramway terminal, sadly demolished as unsafe in 1980, was also built at this time.

Electrification of the tramway was considered by the Tramway Company in 1897, when over one and a half million passengers were carried, following the completion of the last piece of double track.

The Isle of Man Tramways & Electric Power Co. had been heavily reliant on Dumbells Bank to finance construction of the tramway and the cable car system, so when Dumbells collapsed in 1900 the Tramway Company went into liquidation. In September 1901 the Manx Chancery accepted the £50,000 offer from Douglas Corporation for the Horse Tramway and cable car system. Thirty-six cars now operated on the Horse Tramway, 13 double-deckers, three single-deck saloons, 14 open

Tram 18 started life in South Shields as a double-decker, was sold to the Isle of Man in 1887, converted to single-deck in the early 1900s and back to double-deck in 1989. The tram is seen here outside Derby Castle depot, awaiting departure for the Sea Terminal.

The first No. 1 only lasted a few years and was replaced by this enclosed car, seen here between duties at Derby Castle.

toastracks and six roofed toastracks, motive power being provided by 68 horses. The electric tramway was sold in 1902 to a syndicate based in Manchester and later that year to the Manx Electric Railway Company.

In 1906 the Manx Electric Railway Company approached the corporation with the intent to electrify the Horse Tramway in order to provide a through service from Ramsey to Victoria Pier, but this was again rejected, as was a further approach in 1908. Two more tramcars were delivered in 1907 and in 1909 it was ruled that a maximum of eight return journeys were to be operated by each horse in the course of a day.

During the First World War, a winter schedule operated, as the holiday industry slumped. However, by 1920 business was picking up again, with 44 cars available.

Motor buses were introduced along the Promenade for the first time in 1926 and it was proposed that they would eventually take over from the tramway. In 1927, after almost fifty years unbroken service, the tramway began closing for the winter.

Serious threats of closure were hanging over the line by 1933 but the Corporation embarked on a massive track relaying programme, using 65 lb/yd rail which

would last 40 years, thus assuring the tramway of a future. Indeed, the Horse Tramway made a profit of £8,000 in 1933 compared with a loss of £4,280 on the bus services, a pattern that continued for many years. 1935 saw the addition of the offices above the depot at Derby Castle, three more tramcars were added and roller bearings were fitted to the existing fleet. Horses now numbered 135.

The Second World War caused the tramway to close. All the horses were sold, the tramcars put into store, barbed wire erected between the tracks and many sea front hotels requisitioned for prisoners of war.

In April 1946, 42 horses arrived from Ireland, the tramcars emerged from store and a re-opening ceremony was performed on 22nd May by Sir Geoffrey Bromet, the Lieutenant-Governor of the Island. A reduced service was operated but a profit still showed.

Holidaymakers poured back to the Island in 1947, tracklaying restarted and all looked good for the future until 1949, when it was suggested that open top buses should replace the trams, but fortunately the Corporation took no action in this direction. All but one of the double-deck cars had by now disappeared and in 1955 the sole survivor (No. 14) left the island for preservation in the Museum of British Transport at Clapham.

Anniversary celebrations were held in 1956 to mark 80 years service, the horses being paraded at Victoria Pier. The Promenade was lined three or four deep for the special cars, carrying the official party towards Derby Castle.

Holiday traffic declined towards the end of the 1950s and threats of closure were rumoured, but denied officially. Tynwald granted permission in 1961 for a fare increase, as the tramway was now regarded as a speciality ride. The Fleetwood steamer services ceased after the summer of 1961, resulting in a reduction in the number of day trippers to the Island in 1962.

Tynwald Day 1964 saw Her Majesty Queen Elizabeth, the Queen Mother, ride on specially prepared tram No. 44, from Summer Hill to the Villa Marina. The stud of horses was 56 at the beginning of the year being reduced to 43 by the winter. Fifteen new horses were purchased at the beginning of 1965, a year that saw another Royal visit, this time by HRH Princess Margaret and Lord Snowdon.

A flat fare system was introduced in 1966, and the number of passengers showed an increase, which was to continue over the next few years.

HRH the Duke of Edinburgh visited the line in 1970, returning again in 1972, with Her Majesty the Queen, HRH Princess Anne, and Admiral of the Fleet, Earl Mountbatten of Burma. The Royal party rode from the new terminus at Victoria Pier to the Sefton Hotel in cars 44 and 36, both specially painted and decked out with flowers. The Promenade was crowded with a high percentage of the Islanders.

During the early 1970s the Isle of Man Tourist Board had been actively trying to increase the number of visitors to the Island through advertising, which in turn had increased the number of passengers on the Horse Tramway, and by 1973 the numbers were heading towards those of the early 1920s. In 1974 over one and a half million passengers were carried. The price of horses rose sharply during 1974/75 resulting in the Corporation deciding to start a breeding programme, new foals not working the tramway until they were at least three years old.

No. 18, as a single-decker, waits in the rain while its driver enjoys a warm cup of tea.

The centenary fell on 7th August 1976 but as this was a Saturday the celebrations were held on Monday 9th August, by chance during the author's first visit to the Island. The Science Museum in London, by now the home of double-decker No. 14, was persuaded to allow it to return to the Island, be repainted, and take part in the celebrations. Three other cars were also repainted including No. 44 which was only returned to the depot the evening before. Stable staff worked through the night grooming the 50 horses, all brasswork was meticulously polished and nameboards provided for each horse.

At 10 am the horses were led along the promenade towards Victoria Pier where the trams were waiting, having been hauled there five at a time by a double-decker bus. Then at 11.15 the double-deck tram left Victoria Pier at the head of the procession which included the newly restored cable car. Every serviceable tram followed, their passengers being issued with special centenary tickets. It is estimated that 30,000 people lined the Promenade from end to end.

Since the centenary the tramway has seen little change. Some of the restored cars have formed a small exhibition area in the depot at Derby Castle and the total fleet of serviceable cars now stands at 19. The line made a loss in 1981 for the first time in its history, but its future seems assured as one of the leading tourist attractions on the Island. The journey now takes about 20 minutes, at an average of around 6 mph.

Many visitors to the Island enquire about the welfare of the motive power, the horses. The truth is that they are extremely well looked after, spending the winter months grazing in fields around the Island specially reserved for them. During the operating season they are stabled near the Derby Castle terminus and only work four return trips each, in any one day. Indeed, it is claimed that they can count and then stop automatically at the stables at the end of their stint!

Double-decker No. 14, still carrying its centenary year livery, stands in the depot at Derby Castle, where a small display has been set up. This tram has since been moved to the Manx Museum in Douglas.

One of the museum displays built to look like the side of a tramcar and housing many interesting exhibits and old photographs.

Cable car No. 72/73 emerges from the depot prior to working an Isle of Man Railway Society special in May 1991. The tram shop, rebuilt from horse tram No. 22, stands in the adjacent track ready to be hauled out for the day's trading.

41
JACOB'S
DOUGLAS
CORPORATION TRANSPORT

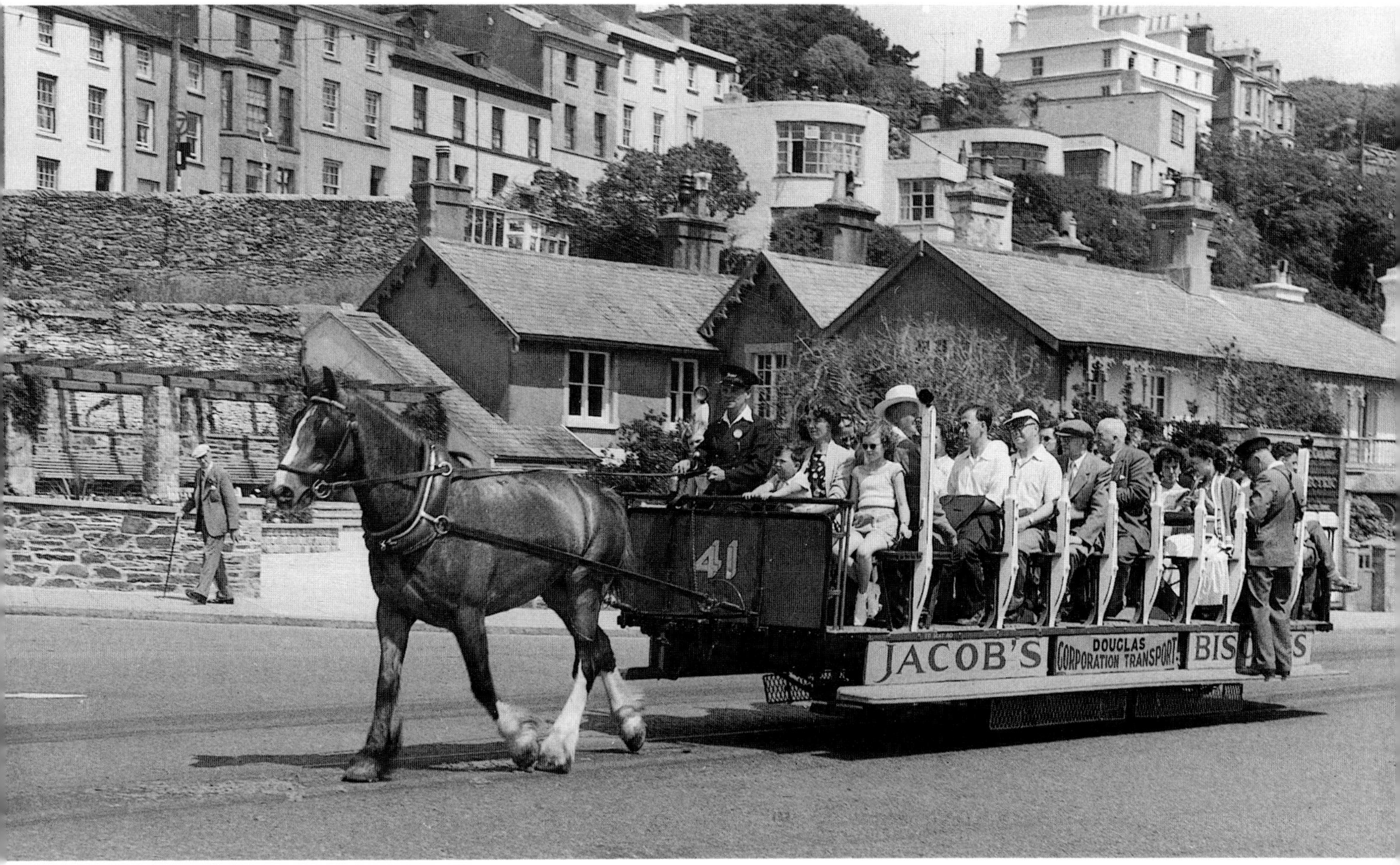

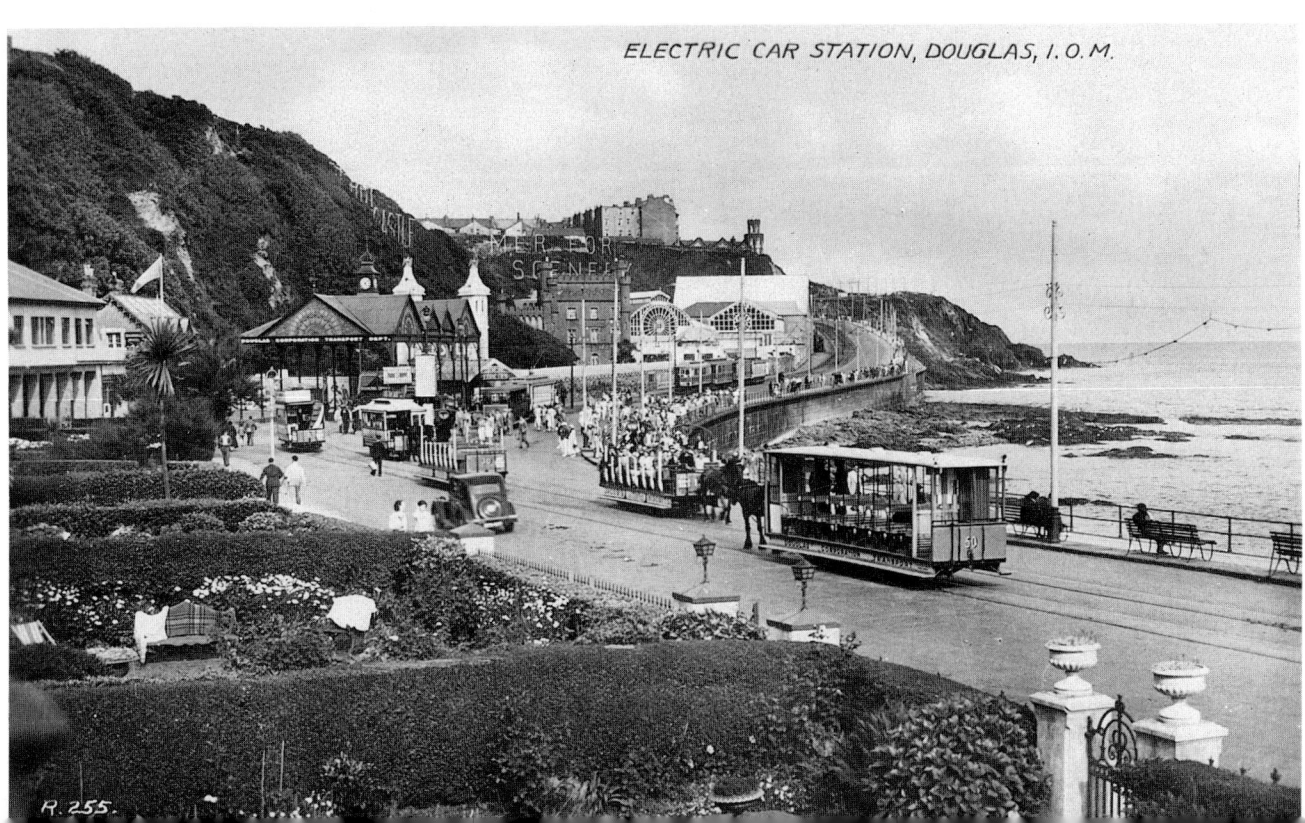

Horse tram No. 41 sets off along the Promenade, past the houses of Summer Hill, during the summer of 1956. Apart from the costume, this scene has hardly changed, but this tram has since been withdrawn.

Vera Taylor

A scene from the past with no less than five horse cars and six electric cars at Derby Castle. The nearest horse car is No. 50, in front of it is No. 30 and in front of that a double-decker, No. 5. The cars on the other track are unidentified, as is the date the photograph was taken.

Author's collection

ELECTRIC CAR STATION, DOUGLAS, I.O.M.
R. 255.

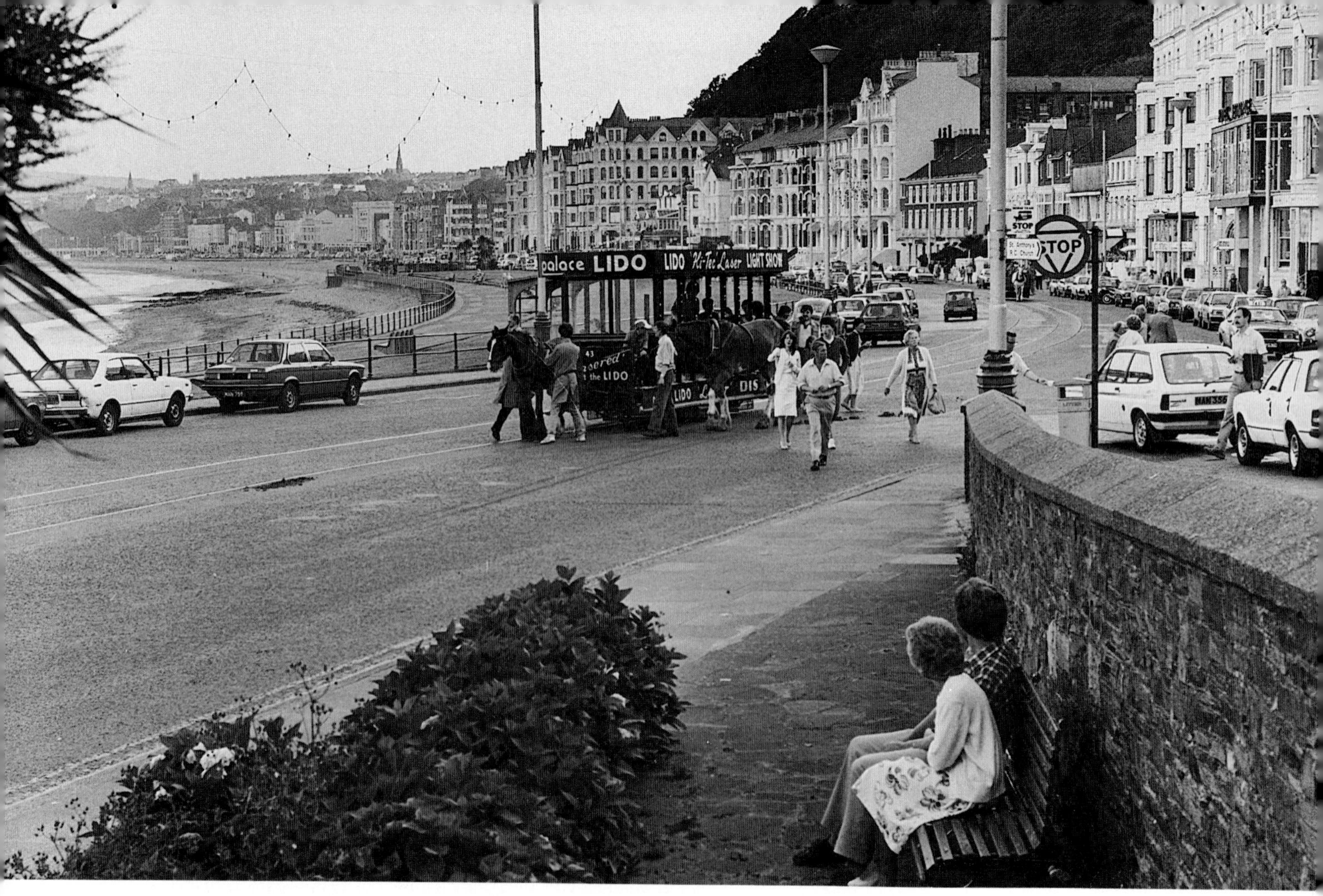
palace LIDO
LIDO
'Hi-Tec Laser' LIGHT SHOW
STOP

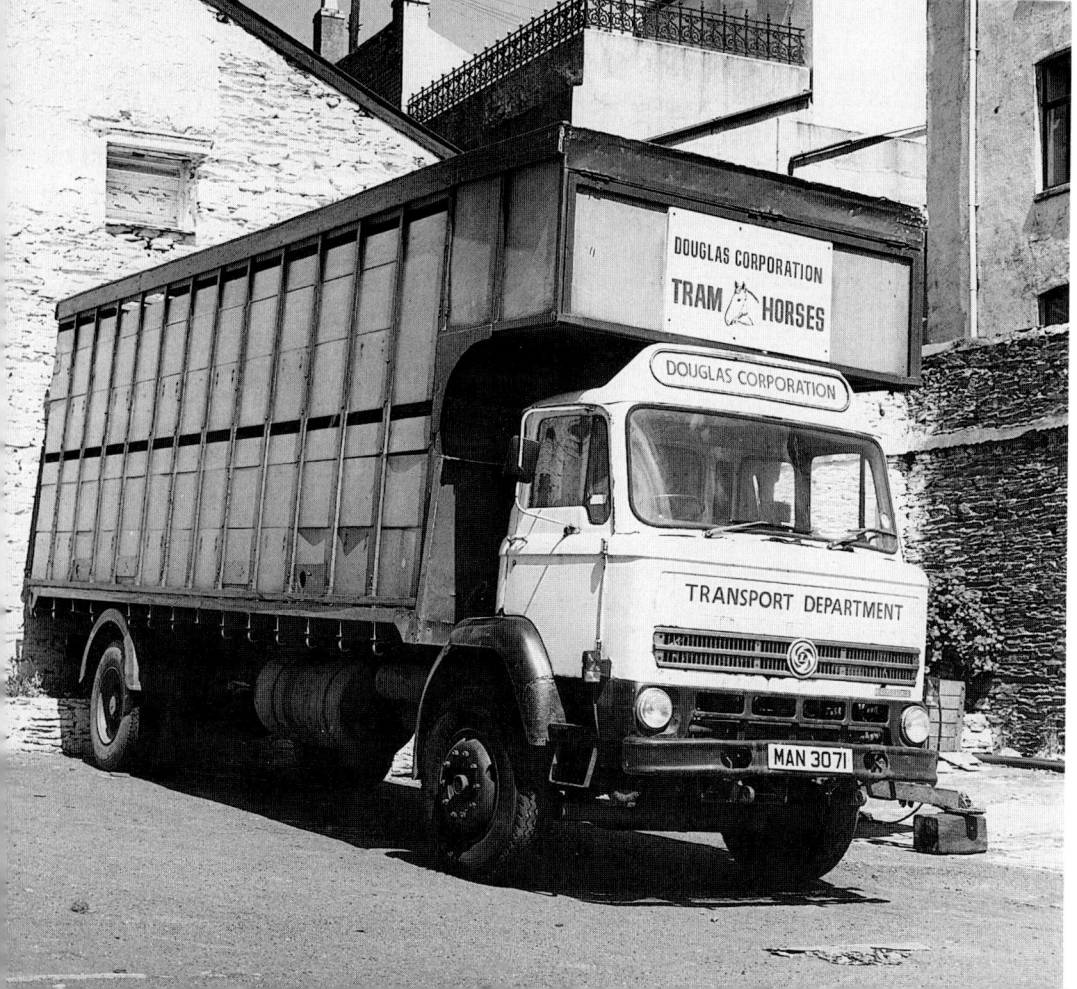

The stables are situated at the foot of Summer Hill and trams stop here to change horses. The next horse is led out as the previous one is uncoupled from tram No. 43 for a well earned rest.

DOUGLAS CORPORATION
TRAM HORSES
DOUGLAS CORPORATION
TRANSPORT DEPARTMENT
MAN 3071

In order to transport the horses to and from their winter grazing grounds, this horse box is employed. It is seen here parked outside the stables.

No. 32 passes the Metropole Hotel whilst heading towards the Sea Terminal. Good use is made of the trams to advertise local attractions.

The double-deck car No. 18 makes one or two trips each day, a supplementary fare being payable. Here it passes the Castle Mona Hotel on its way to the Sea Terminal.

Island Photographics

The Falcon Cliff Hotel stands high on the hillside above the Douglas promenade. The original cliff lift, built in 1887, was closed in 1896 and moved to Port Soderick. A replacement was built by Wadsworth & Sons of Bolton in 1927, slightly to the south of the original. One car was provided, operating on a 5ft gauge track, with the counter-weight between the rails. The hotel and lift were owned by Okell's Brewery of Douglas but have recently been offered for sale; the future of the lift is therefore uncertain.

A well-loaded tram No. 32 heads towards Derby Castle. The traffic arrows, seen here on the end panel of the tram, are recent additions.

Island Photographics

The Douglas Cable Car Group's immaculate preserved car is seen being propelled past the Balmoral Hotel by a Land-Rover, during a special trip for the Isle of Man Railway Society in May 1991. One wonders why the builders went to the trouble of fitting opening vents along the clerestory when the car is open-sided anyway!

No. 18, as a single-decker, hurries towards the Sea Terminal on a cold damp day in May 1986. Derby Castle station can just be seen over the horse's back.

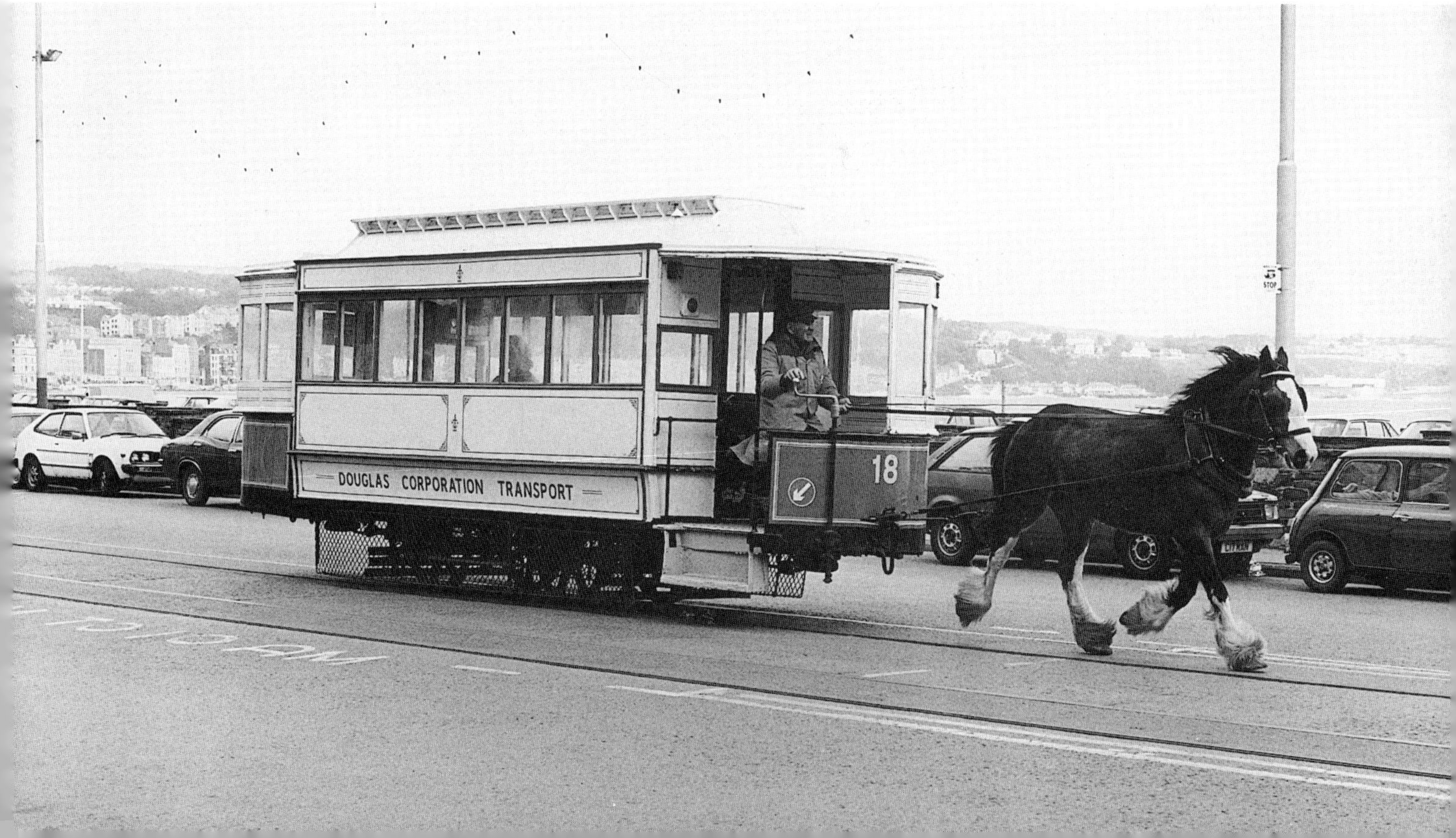

LOCH PROMENADE, DOUGLAS, I.O.M.

This pre-1938 scene shows open tram No. 39 passing No. 44 whilst on its way to the Sea Terminal. The sign on the clock reads: 'Last cars leave at 11.20 to the Promenades and Bucks Road'. Bucks Road was on the cable car route, the horse trams serving the Promenades.

Author's collection

Tram No. 39, photographed some fifty years later, passes the same spot as in the previous picture. Notice how the wire hoops have given way to wooden advertising boards over the end seats.

The clock has recently been restored and repainted. The road to the left formed part of the cable car route and a piece of preserved track is laid in the island on which the clock stands. Tram No. 37 nears its terminus.

Testing the cable cars in 1898. Here, No. 72, one of the two used in the preserved car, ascends Prospect Hill.

Horse tram No. 1 arrives at the Sea Terminal and hence the end of the line, the return crossover being clearly visible. The trams used to continue on to Victoria Pier but this would involve crossing a busy road adjacent to a roundabout.

Manx Museum and National Trust

The Steam Railway

The first attempts to build a railway from Douglas to Peel were made in 1860 and 1863 but neither project came to anything, nor did a scheme registered with Tynwald in 1865. Eventually, in April 1870, a meeting was called in Douglas, to which all interested parties were invited, to come up with a plan that would not suffer the fate of previous attempts.

Agreement was reached in principle to form a railway company to build a line linking Douglas with Peel, Ramsey and Castletown, with a projected extension to Port Erin which was to have a steamer service to Holyhead in North Wales.

The Isle of Man Railway Company was registered in December 1870 with a capital of £200,000 but, following a survey of the routes, it was realised that this was not sufficient. Approaches were made overseas and, after a short while, the company was joined by The Duke of Sutherland and Mr John Pender. The Duke became Chairman for the first seven years, being followed by Mr Pender.

In 1872 tenders were invited to construct a 3ft gauge railway from Douglas to Peel and Port Erin. Four replies were received and the contract was awarded to Messrs Watson & Smith of London.

Three locomotives were ordered from Beyer, Peacock of Gorton Foundry, Manchester in 1872. They were 2-4-0T locomotives with copper capped chimneys, brass domes and brass numerals on the chimneys. Eventually 15 such locomotives were delivered to the railway, a decision that proved useful in later years as it was possible to interchange most parts between locomotives. The livery was to be green with a black smokebox.

No. 1 *Sutherland* arrived in March 1873 and Nos 2 and 3, *Derby* and *Pender*, in June. Rolling stock was provided by the Metropolitan Carriage & Wagon Co. of Saltley consisting of 29 4-wheeled coaches including two guards vans. Various items of freight stock also arrived.

The first train ran on 1st May 1873 with the Duke of Sutherland on the footplate of No. 1, the train formed of open wagons with seating fitted. No. 1 came off the rails at Peel, the contractor's locomotive being summoned to return the train to Douglas. The Duke returned to the railway on 1st July 1873 to take his seat in the directors' saloon forming part of the twelve-coach official opening train. Locomotive No. 1 hauled the train carrying suitable decoration and a banner reading 'Douglas and Peel United'. Huge crowds turned out to welcome the railway to the Island and to watch the train clatter past at speeds of up to 25 mph, faster than any Islander had seen before.

Meanwhile, over 400 men were busy pushing on with the Port Erin line which was proving a more difficult task with its steep gradients and sharper curves. In June 1874 the contractor pulled out, leaving the railway company to complete the line, which in the end cost £9,875 per mile to build. This meant that economies were made on stations with only simple wooden structures being provided. The line opened without ceremony on 1st August 1874.

Two further locomotives, Nos 4 and 5, *Loch* and *Mona* were delivered in 1874 to work the new line and the rolling stock fleet was increased to 56 coaches, still all 4-wheeled. Locomotives Nos 6 and 7, *Peveril* and *Tynwald*, were delivered in 1875 and 1880 respectively.

The residents in the north of the Island thought they would be next to get a railway but the company announced that it had no plans for such an extension. The northerners decided to go it alone and the Manx Northern Railway Company was formed in 1877, and given government consent in 1878 to build a 16-mile line from St Johns to Ramsey.

The contract to build the line, which was to include

11 pm in Douglas shed! No. 12 *Hutchinson*, with No. 4 *Loch* just visible behind, stands beside a third locomotive. Note the firelighting wood, salvaged from old sleepers, stacked in the cab of the locomotive on the right, awaiting the arrival of the morning shift.

several viaducts, was awarded to J. & W. Grainger of Glasgow, with completion due by 1st July 1879. Two locomotives were ordered from Sharp, Stewart & Company of Manchester; numbered 1 and 2 they were named *Ramsey* and *Northern* respectively. Coaches were all 6-wheelers. A third locomotive, No. 3 *Thornhill* was ordered for delivery in 1880, this time from Beyer, Peacock and similar to those being supplied to the Isle of Man Railway Company.

Two and a half miles south of St Johns, the small mining village of Foxdale was producing 4,000 tons of ore annually and the idea of a railway linking Foxdale with the other lines at St Johns was actively considered. The Foxdale Railway Company was formed in 1882, managing to lease itself to the Manx Northern Railway for 50 years, even before construction started.

The $2\frac{1}{2}$ mile line, with stations at Foxdale and St Johns, was constructed by Hugh Kennedy & Sons of Glasgow. No expense was spared and by June 1886 the line was in full swing, carrying tons of ore each day. A fourth locomotive was delivered, this time an 0-6-0T from Dubs & Co., Scotland, No. 4 *Caledonia*.

The final years of the 1800s saw the Isle of Man Railway go from strength to strength, 26 bogie coaches being delivered by 1896 and locomotives No. 8 *Fenella* and No. 9 *Douglas* arriving in 1894 and 1896. Douglas station was rebuilt with the impressive forecourt that still exists today, the carriage shed was built and extensions were added to the locomotive sheds and workshops.

Port Erin, Port St Mary, Peel and Castletown stations were also rebuilt, as was that at Port Soderick where a passing loop was added. The tourist boom had not gone unnoticed by other railway promoters and the opening of the Derby Castle to Groudle electric railway, which eventually reached Ramsey in 1899, provided some interesting competition. The Isle of Man Railway Company advertised a 68-minute journey from Douglas to Ramsey, while the electric railway offered 70 minutes.

The Foxdale Railway went into voluntary liquidation in 1891, leaving the Manx Northern to carry out the

The sheds at Douglas are divided into three areas: the steam running shed, the locomotive workshop and the carriage workshops. No. 4 *Loch* is in the running shed, being prepared for the following day's work, after arriving from Port Erin with the last train of the day.

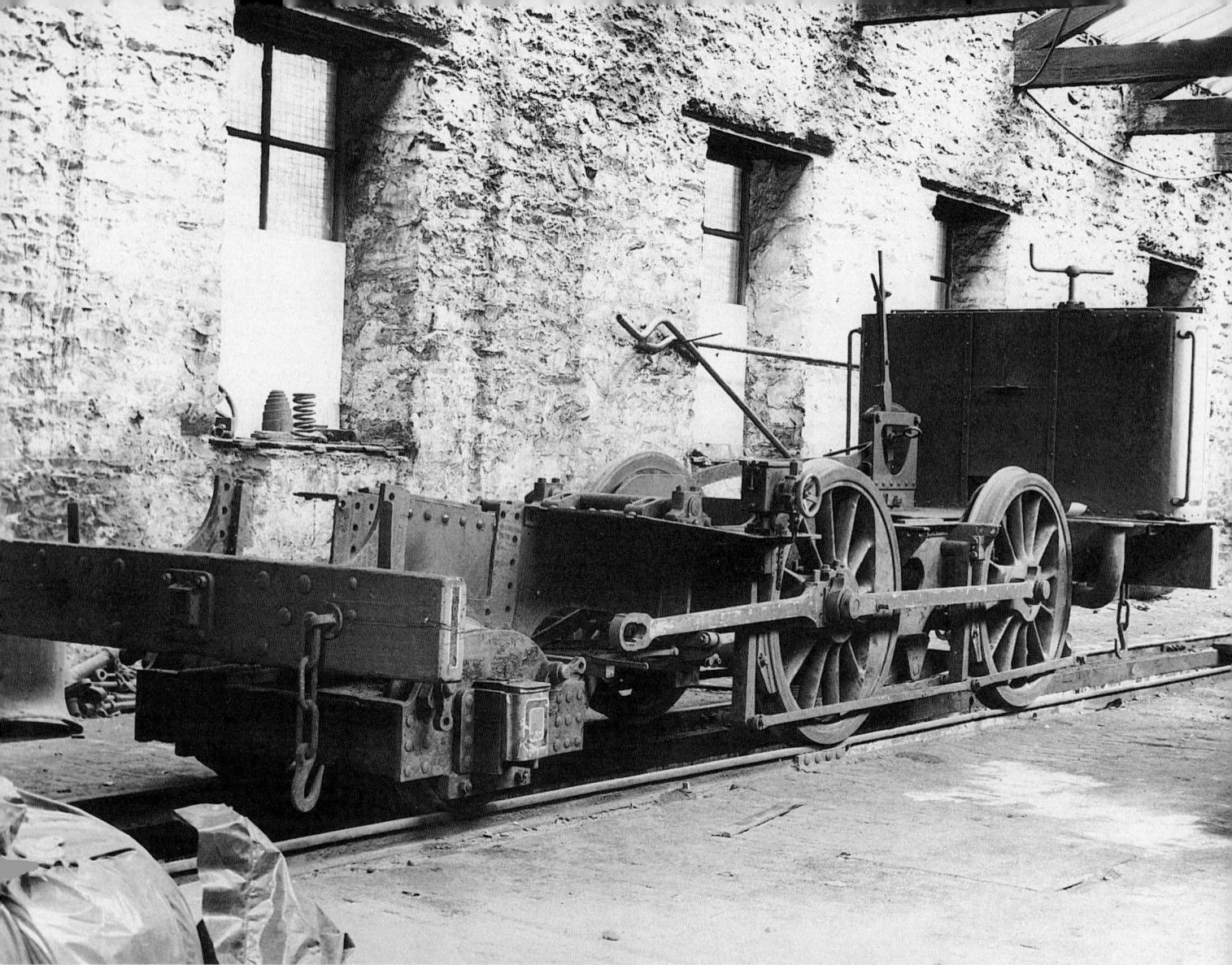

Standing in the back of the running shed are the frames and wheels of No. 10 *G.H. Wood*. This chassis is now in the adjacent workshop being rebuilt, ready to receive the boiler from No. 13 *Kissack*. The 'new' No. 10 should be ready for service in 1993.

terms of the lease, until in 1904 the Manx Northern was itself in trouble. The Isle of Man Railway was approached and an Act of Tynwald allowed the purchase of the Manx Northern for £60,000 and the Foxdale Railway for £7,000.

New coaches were ordered, the four Manx Northern Railway locomotives were taken into Isle of Man Railway stock and two more new locomotives arrived from Beyer, Peacock in 1905, No. 10 *G.H. Wood* and No. 11 *Maitland*.

A further two locomotives arrived from Beyer, Peacock in 1908 and 1910, numbered 12 and 13 and named *Hutchinson* and *Kissack*. Port Erin station was enlarged in 1911, Peel was rebuilt and the canopies were added at Douglas. One million passengers were carried for the first time in 1913. The Foxdale mines closed in 1914 but the passenger service continued, usually with just one coach.

The outbreak of war in 1914 reduced the number of visitors but, from 1915, the railway was called upon to operate trains between Peel and Knockaloe over a specially constructed line serving a camp for 20,000 prisoners of war. The former Manx Northern locomotive *Caledonia* worked the line as it was the best to cope with the steep gradients.

Visitors returned to the Island by the thousand after the war and, despite the rising costs of materials and wages, and the introduction of the eight-hour working day, the railway continued to flourish. Much overtime was worked by the staff to keep the frequent service operating and, during the winters, all attention was turned to maintenance of locomotives, rolling stock and track.

No. 16 *Mannin* was delivered in 1926 and was the last locomotive ordered new by the railway and the last new coach, No. F49 also arrived. Manxland Bus Services were granted permission to operate motor bus services on a range of routes from Douglas. The railway response was to cut fares, reduce journey times and increase the number of trains timetabled, particularly during the evening.

The early bus years culminated in the formation of Isle of Man Road Services Ltd in 1930, resulting in co-operation between the bus and rail operators, 'Freedom of the Island' tickets being offered jointly. The latter years of the 1930s saw an annual rail passenger level of around 750,000.

World War II brought more work for the railway but,

as the Knockaloe camp had been demolished, hotels were commandeered to house prisoners of war. Training camps at Jurby and Castletown took passenger numbers above those of peacetime and school journeys came back from the buses because of shortages of tyres and fuel. Over 14,000 special trains had operated by the end of hostilities in 1945.

Visitors once again returned in large numbers in 1946. However, the steam locomotives were all in need of overhaul, several requiring complete new boilers, of which three were ordered and fitted. During 1945, No. 7 had been cannibalised for much-needed spares and No. 2 went the same way in 1951. Nos 3, 4 and 9 were also withdrawn from service, in need of new boilers.

The post-war boom ended in 1956 with over a million passengers carried for the last time, the numbers steadily declining thereafter as air travel took holidaymakers off to other parts of the world. Costs continued to rise,

receipts fell and service cuts started. Sunday services were withdrawn altogether and winter services were reduced to school traffic level without much loss of receipts.

No. 11 *Maitland* received a badly needed new boiler in 1959 but replacements for Nos 1, 6, 13 and 16 were considered unnecessary. Service cuts continued and winter services became virtually non-existent.

Two former County Donegal diesel railcars were obtained in 1961, at £160 each, and were refurbished in time to work the Peel service that winter. The situation on the railway was now serious and the outlook bleak.

On 7th July 1963 Her Majesty Queen Elizabeth the Queen Mother arrived on the Island and travelled on the railway from Douglas to Kirk Braddan in coach F36, No. 11 doing the honours at the front.

A loss of £8,000 was made in 1965 and all winter services were cancelled in order to carry out much-needed maintenance; this too was cancelled without warning. No

The removal of the boiler from No. 13 *Kissack* in early 1992 revealed the poor condition of the other parts. This locomotive will now be rebuilt gradually and hopefully, in due course, a new boiler will be acquired, but return to service is unlikely to be within five years.

trains ran in 1966 but a report into the future of the line recognised the importance of the railway to the community. The Isle of Man Railway Supporters Association was formed, offering to assist in any way it could. Relief arrived from the Marquis of Ailsa, who agreed to lease the line for 21 years with an opt-out clause after five years.

Hundreds of people turned out on 3rd June 1967, the occasion of the railway's re-opening. Five special trains ran to Peel, with services to Ramsey and Castletown as well and a frequent daytime service operated throughout the railway for the rest of that season.

No. 4 *Loch* received a new boiler in 1968, returning to service on the last day of that season. However, mounting losses again caused the season to be cut short, the last train from Ramsey running on 6th September and the last from Peel the next day. The Peel and Ramsey lines never re-opened.

The Tourist Board agreed in 1968 to assist in keeping the Port Erin line open for three years. No. 13 received a new boiler and passenger numbers increased slightly but the Marquis was still losing money and announced that he would take the five year option. It is estimated that he lost around £43,000 during his stay but, without him, the railway probably would not have survived.

In 1971 the Tourist Board offered the railway company further support for another three years and costs were reduced by only operating a Monday to Friday service. Her Majesty The Queen travelled on the line from Castletown to Douglas on 2nd August 1972, the total number of passengers carried being the best for some time.

On 1st July 1973 the centenary of the opening of the Peel line was marked by a special train on the Port Erin line, which celebrated its own centenary in style the following year. On 1st August 1974 No. 4 *Loch*, also celebrating 100 years and having completed 2,000,000 miles, hauled a special train from Douglas, stopping at every station to pick up guests to attend a fete in Port Erin.

Tynwald debated the future of the line in 1974 and, after much discussion, the railway company was offered assistance to run four trains daily except Saturdays, between Port Erin and Castletown from 1975 onwards. These were extended to Ballasalla in 1976, pressure now being concentrated on a return to Douglas, still the main tourist centre. Government approval came in 1977, the same year as the Manx Electric returned to Ramsey, so the entire Port Erin line was open once again.

The Government was persuaded to purchase the line in 1977, following the election of new members at the Manx general election of 1976. The steam railway became part of the Manx Electric Railway Board's responsibility, under the title 'Isle of Man Railways'.

The canopies at Douglas station were removed and Port Soderick station building was sold, as part of a drive to reduce costs and tidy up the railway. Douglas station also lost its signals and the eleven tracks were reduced to five to reduce loan costs. At Ballasalla, surplus land was sold for office development in return for a new station building on the other side of the line. Santa specials were introduced and now form an important part of the railway calendar. New boilers were obtained for Nos 11 and 12, extensive track repairs carried out and coaches renovated.

In 1983 the Manx Electric Railway Board became The Isle of Man Passenger Transport Board and this in turn became part of the Department of Tourism and Transport in 1986.

The railway has continued to operate each season since, the intervening winters being used to improve further the quality of track and stations and to tidy up the lineside. Locomotives are given regular overhauls and more coaches are being thoroughly refurbished. The line is now in better condition than for many years and the mid-summer service has been increased to six trains each way daily, to cope with demand.

The railway celebrates 120 years of service in 1993 and can now look forward to its 150th anniversary with confidence.

Much of the engineering required to keep the steam locomotives running is carried out at Douglas. This wheel lathe, loaded with a set of locomotive driving wheels, is just one of the many machines that line the workshop walls.

No. 12 *Hutchinson* undergoes overhaul in the workshops at Douglas. The small overhead crane can be seen behind the locomotive, which is standing on accommodation skates to allow movement within the workshop. The locomotive's driving wheels are visible behind the locomotive.

This July 1979 picture of No. 4 *Loch* shows the painted-on builder's 'plate', with which it ran for a while in the late 1970s. The now-replaced semaphore signals are visible through the locomotive's smoke.

The impressive entrance to Douglas station, during refurbishment in July 1984. The glass has now been replaced in the canopy and the interior restored to excellent standards, incorporating a cafeteria.

No. 11 *Maitland* awaiting departure with the first southbound train of the day, the 10.10 to Port Erin. The train will take one hour and five minutes to reach its destination, just under $15\frac{1}{2}$ miles away.

As the first train is at 10.10 and the last at 16.10, trains running in the dark are rare. However, the author was lucky to be on the Island when No. 4 *Loch* worked a 23.15 special to Port Erin on 27th May 1989, which was duly photographed prior to its departure from Douglas.

With a combined age in excess of 170 years, Nos 12 and 13 prepare for the day's work at Douglas. The running shed, visible between the two locomotives, appears to be occupied by a coach. During 1992 extensive renovation work was carried out to the sheds.

During the summer months steam locomotives can cause lineside fires and the Isle of Man is no exception. The chassis of coach F23, built in 1896, had its bodywork removed in 1983 and was fitted with two water tanks for use as the fire train. After the arrival of the last train it is not unusual for this vehicle to be taken down the line to extinguish small fires. It is seen here being shunted by No. 12 *Hutchinson* while No. 4 *Loch* awaits its next duty in the loop.

Arriving at Douglas in 1979. The semaphore signals have since disappeared and the signal box is disused, although preserved. The building on the right is part of the depot, the carriage shed being behind the signal box.

No. 12 *Hutchinson* leaves Douglas for Port Erin with five coaches in tow. The track on the right leads to the carriage shed, which is the building to the right of the picture. A withdrawn Leyland National bus stands alongside the shed.

After leaving Douglas the line climbs steeply through woods. No. 11 *Maitland* sets off up the bank with four coaches for Port Erin. Note the use of former British Rail sleepers to form the coal yard fence.

Having climbed out of Douglas, the line runs beside the coastal road above Port Soderick. No. 11 *Maitland* heads south with a well-loaded train for Port Erin.

Left: During a brief visit to the Island in 1992 the author visited the A25 road overbridge at Oakhill, on the outskirts of Douglas, to photograph the 14.10 to Port Erin. Much to his surprise No. 11 appeared hauling eight coaches, banked by No. 12. The entire train worked through to Port Erin and back. The recently relaid track is clearly evident.

No. 11 pauses at Port Soderick with the 14.10 from Douglas. The station building is now privately owned.

Santon is the second station on the line, just over $5\frac{1}{2}$ miles from Douglas and is a request stop. Any passenger wishing to board must give a hand signal to the driver. Here, No. 12 draws to a halt in July 1984.

David Masters

Santon station building was restored in 1982 with funds provided by the Isle of Man Railway Society but it was beginning to look weather-beaten again by the time this picture was taken in July 1985.

No. 11 *Maitland* runs through Santon, past the siding containing three flat rail wagons, one van and an open wagon. This Douglas-bound train is the Sunday to Thursday 15.30 from Port Erin.

The railway purchased two former County Donegal railcars in 1961. They are numbered 19 and 20 and are seen here approaching Santon station. The photographer had signalled to the driver that he wished to join the train but it was already overfull and failed to stop, leaving him to travel back to Douglas on a flangeless vehicle! The frames of No. 7 *Tynwald* are seen on a flat wagon to the left of the picture.

David Masters

Ballasalla is the main passing point for the normal service, the train from the south often arriving before the one from the north. Here No. 12 waits patiently for No. 4 to arrive from the north, before continuing its journey to Douglas.

A telephoto shot of No. 4 arriving at Ballasalla's new station with a southbound train. This brick building and raised platform, on the 'down' side, replaced the old wooden building on the 'up' side.

Island Photographics

Ballasalla station viewed from the level crossing at the south end. The site of the original station buildings is now occupied by an office block and a car park, while passengers enjoy the luxury of the new facilities.

The Ballasalla water tower situated at the southern end of the station on the 'down' side.

The new station building at Ballasalla, photographed in 1985 during its first season of service.

Van No. G19 stands in a siding at Ballasalla. This van, built at Douglas in 1921 on the chassis of van E3, is fitted with roof platforms for tree lopping and has a vacuum brake pipe for operating between the diesel railcars.

Open wagon M70 labelled 'Not to be moved' stands at Ballasalla in 1981. Supplied by the Metropolitan Carriage & Wagon Co. in 1925, it is seen here in use as a sleeper store.

After leaving Ballasalla the next major stop is at Castletown but Ronaldsway Halt lies between these two stations. After taking this photograph in May 1991, the author walked the few yards to the nearby Ronaldsway airport and flew back to London. Perhaps in due course a full scale station could be developed here, or even a branch into the airport.

The steam locomotives all carry brass name-plates of this style. This example belongs to No. 12 and has a black background, while some have a red background.

Each locomotive also carries a builder's plate, this one belonging to No. 12. This picture also shows to advantage the excellent care that is taken of these fine locomotives, so much polish having been applied to this plate that some of the paint has worn off.

No. 13 *Kissack* stands at Castletown while working to Douglas from Port Erin. The number of passengers waiting on the 'down' side would indicate the imminent arrival of a train for Port Erin, for which *Kissack* must wait before proceeding.

Castletown station building, which replaced the original structure in 1901, and has remained virtually unaltered since. The Isle of Man Railway Society has recently provided funds for the restoration of the wooden canopy.

The chassis of coach F54 stands in the siding at Castletown in May 1986. Supplied by Metropolitan Carriage & Wagon Co. in 1923, it received two 3-compartment, former 4-wheeled bodies dating from 1878. This old bodywork was later scrapped, and the chassis is now in Douglas workshops receiving a new body.

The Castletown stationmaster looks on as No. 11 *Maitland* blackens the skies above his station after arrival from Douglas with the 10.10. It is due to depart from here at 10.56.

No. 11 awaits the right of way from Castletown with a six-coach train. Three or four coaches are normal, five or six usually indicating a booked party.

The driving cab of one of the railcars, the diesel engine being housed in the casing on the left. The cabs, along with the power bogies, were built by Walker Brothers of Wigan in 1949/50. The passenger part of the car, which is articulated to the cab, was built at Dundalk Works.

The former County Donegal railcars approach Castletown with a special working on Spring Bank Holiday Monday 1986. These railcars are currently in need of major overhaul before being used in service again.

No. 4 *Loch* accelerates away from Castletown with a southbound seven-coach train.

The 16.15 Port Erin to Douglas on Monday, 25th May 1992 with No. 12 on the front and No. 11 on the back. This remarkable train is seen having just passed Ballabeg station on its way north.

Ballabeg station, photographed in July 1984, is a request stop. A considerable amount of work has since been carried out here.
David Masters

PETTER
OIL ENGINES
& ELECTRIC LIGHTING PLANTS

This 1992 photograph of Balla-beg station shows the canopy added to the building and the cleared platform area.

No. 11 *Maitland* approaches the road overbridge at the western end of Ballabeg station with the 12.05 Port Erin to Douglas service on 23rd May 1992.

Colby is the next stop on the journey to Port Erin. No. 13 *Kissack* pauses at the station with a southbound train in August 1984.

David Masters

No. 12 leaves Colby and rejoins the single track with an Isle of Man Railway Society special in May 1991. The locomotive would have made short work of the two coaches that made up its train.

Nos 11 and 12 at Colby. *Hutchinson* had arrived with an 09.30 Douglas to Port Erin special and was overtaken by *Maitland* working the 10.10 service train over the same route. The loop at Colby is only likely to see use for events such as this.

Nearing the end of its journey, No. 11 *Maitland* crosses the main A5 road as it approaches Port St Mary. This crossing hut has been renovated recently, the area around it cleared and the track relaid.

The impressive Port St Mary station building, used until recently as a hostel. The stationmaster has been banished to the wooden hut just visible to the left of the picture.

A small running shed is provided at Port Erin, within the buildings that house the museum adjacent to the station. No. 11 has arrived from Douglas and moved into the shed for attention before returning north. One locomotive is stabled here each night.

Carol Edwards

No. 12 stands in the bay platform at Port Erin awaiting departure with the Sunday to Thursday only 15.30 to Douglas. The bay is used for this train because the 14.10 from Douglas will arrive in the main platform.

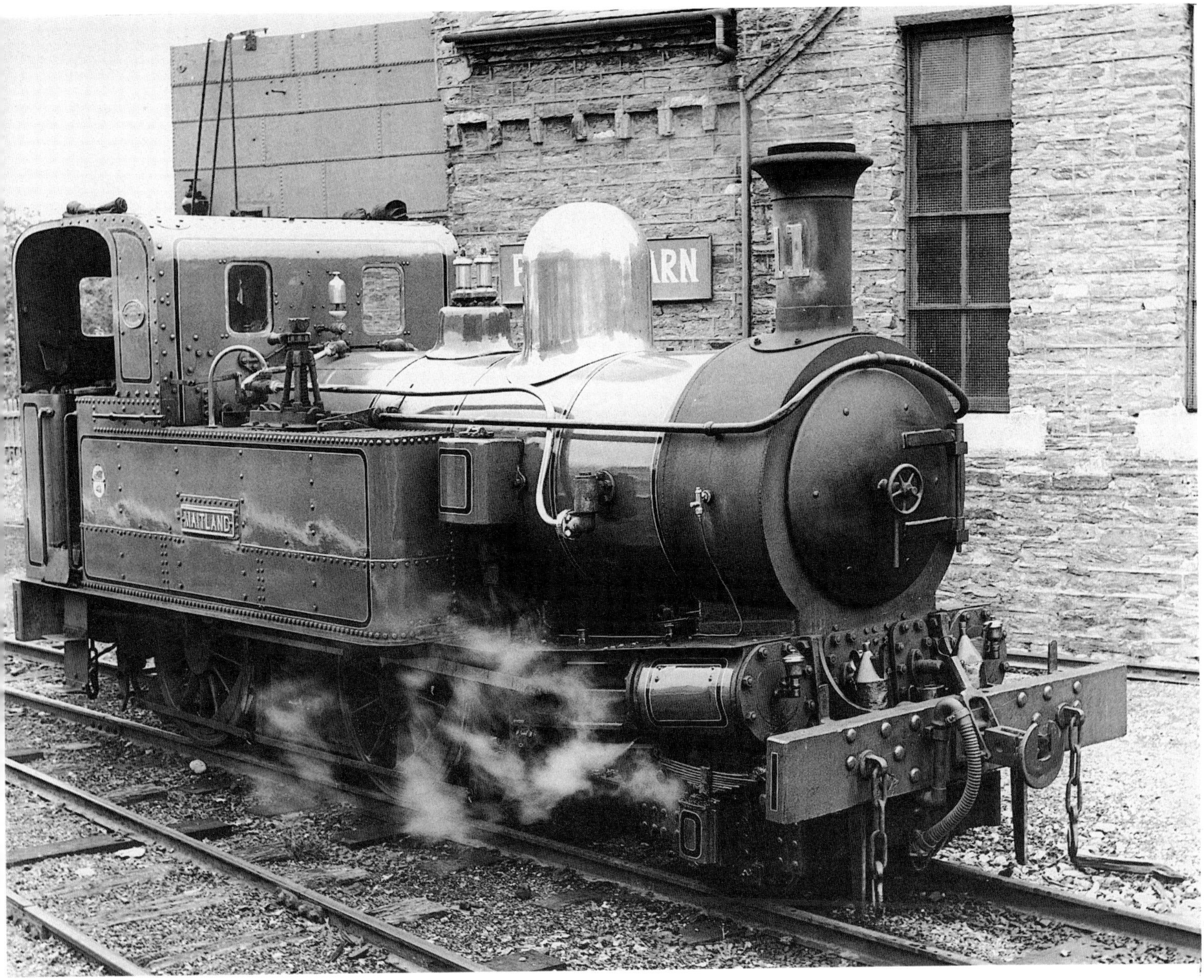

A formal front three-quarter view of No. 11 standing at Port Erin. The superb condition of these locomotives is evident.

The refurbished ticket hall at Port Erin station, with various models in cabinets and disused headboards displayed on the walls. The station Tea Room is through the door on the left.

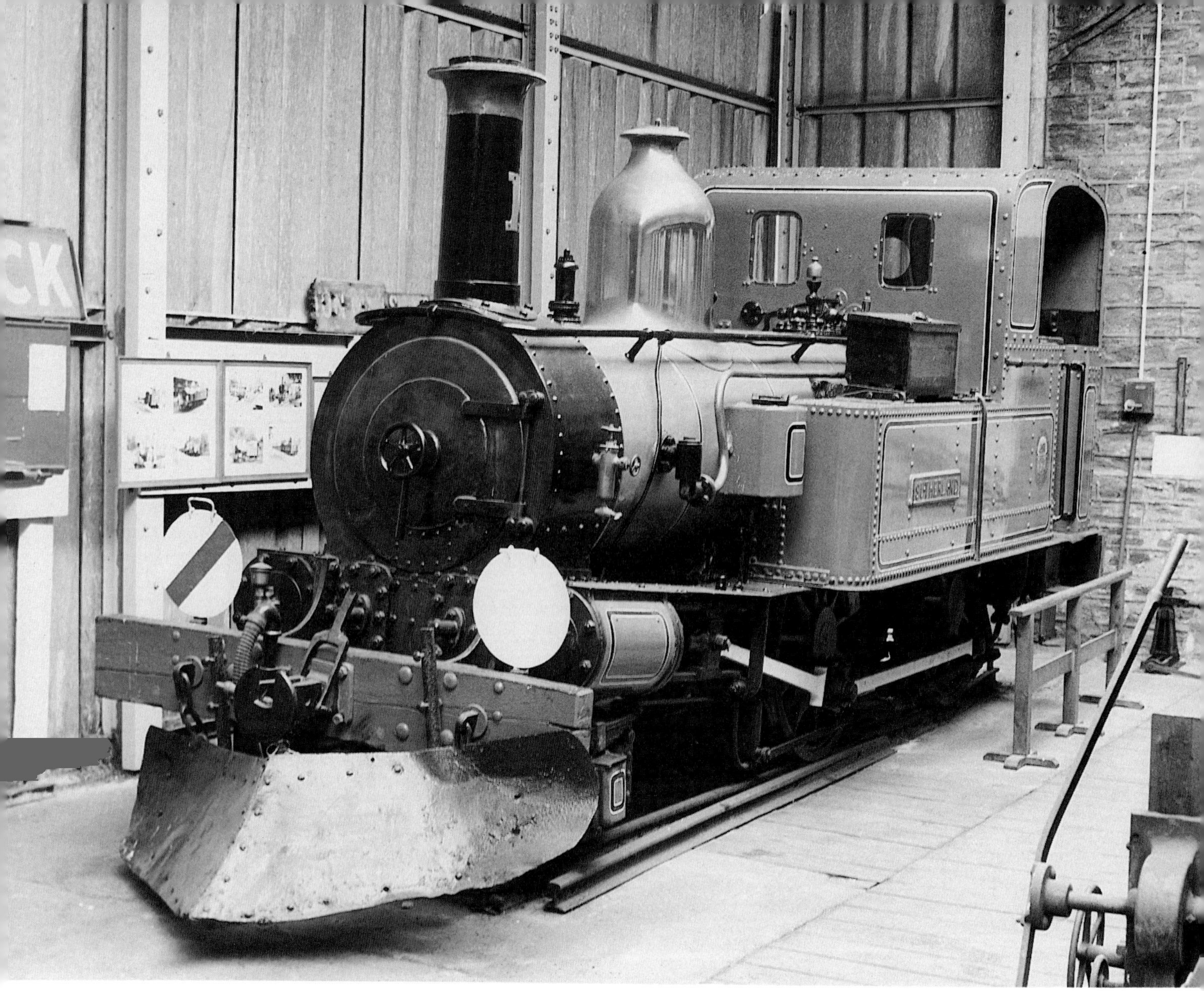

The original Isle of Man Railway locomotive No. 1 *Sutherland*, named after the Duke of Sutherland who was company chairman at the opening of the Peel line in 1873. Perhaps one day this fine locomotive will leave the confines of the museum at Port Erin, and again be seen steaming along the line to Douglas.

No. 9 *Douglas*, built in 1896, is the property of the Isle of Man Railway Society. The society intends to return one of their other locomotives, No. 8 *Fenella* to service but no firm plans appear to exist for No 9, which is in a sorry state, as this picture shows. Note that the numerals on the chimney are still in place.

Manx Electric Railway Fleet List

Power Cars

Car No.	Built by	Year	Bogies by	Motors	Car Type	No. of Seats	Length	Width	Height	Notes
1	Milnes	1893	Milnes [1] S3	2 × 25hp	Unvestibuled saloon	34	34' 9" 10.59m	6' 6" 1.98m	11' 0" 3.35m	In service [2]
2	Milnes	1893	Milnes [1] S3	2 × 25hp	Unvestibuled saloon	34	34' 9" 10.59m	6' 6" 1.98m	11' 0" 3.35m	In service
3	Milnes	1893	Milnes [1] S3	2 × 25hp	Unvestibuled saloon	34	34' 9" 10.59m	6' 6" 1.98m	11' 0" 3.35m	Destroyed Laxey 1930
4	Milnes	1894	Milnes [3] S3	2 × 25hp	Vestibuled saloon	36	34' 8" 10.56m	6' 3" 1.90m	11' 0" 3.35m	Destroyed Laxey 1930
5	Milnes	1894	Milnes [1] S3	2 × 25hp	Vestibuled saloon	32	34' 8" 10.56m	6' 3" 1.90m	11' 0" 3.35m	In service
6	Milnes	1894	Milnes [1] S3	2 × 25hp	Vestibuled saloon	36	34' 8" 10.56m	6' 3" 1.90m	11' 0" 3.35m	In service
7	Milnes	1894	Milnes [1] S3	2 × 25hp	Vestibuled saloon	36	34' 8" 10.56m	6' 3" 1.90m	11' 0" 3.35m	In service
8	Milnes	1894	Milnes [1] S3	2 × 25hp	Vestibuled saloon	36	34' 8" 10.56m	6' 3" 1.90m	11' 0" 3.35m	Destroyed Laxey 1930
9	Milnes	1894	Milnes [1] S3	2 × 25hp	Vestibuled saloon	36	34' 8" 10.56m	6' 3" 1.90m	11' 0" 3.35m	In service
10	Milnes	1895	Milnes S3	2 × 25hp	Vestibuled saloon	46	37' 3" 11.35m	6' 9" 2.05m	10' 0" 3.04m	Withdrawn 1902 [4]
11	Milnes	1895	Milnes S3	2 × 25hp	Vestibuled saloon	46	37' 3" 11.35m	6' 9" 2.05m	10' 0" 3.04m	Withdrawn 1902 [5]
12	Milnes	1895	Milnes S3	2 × 25hp	Vestibuled saloon	46	37' 3" 11.35m	6' 9" 2.05m	10' 0" 3.04m	Withdrawn 1902 [6]
13	Milnes	1895	Milnes S3	2 × 25hp	Vestibuled saloon	46	37' 3" 11.35m	6' 9" 2.05m	10' 0" 3.04	Withdrawn 1902 [7]
14	Milnes	1898	Milnes S3	4 × 20hp	Cross bench open	56	35' 5" 10.79m	6' 3" 1.90m	10' 6" 3.20m	Serviceable
15	Milnes	1898	Milnes S3	4 × 20hp	Cross bench open	56	35' 5" 10.79m	6' 3" 1.90m	10' 6" 3.20m	Stored
16	Milnes	1898	Milnes [8] S3	4 × 20hp	Cross bench open	56	35' 5" 10.79m	6' 3" 1.90m	10' 6" 3.20m	In service
17	Milnes	1898	Milnes S3	4 × 20hp	Cross bench open	56	35' 5" 10.79m	6' 3" 1.90m	10' 6" 3.20m	Stored
18	Milnes	1898	Milnes S3	4 × 20hp	Cross bench open	56	35' 5" 10.79m	6' 3" 1.90m	10' 6" 3.20m	In service
19	Milnes	1899	Milnes [9] S3	4 × 20hp	Winter saloon	48	37' 6" 11.43m	7' 4" 2.23m	11' 0" 3.35m	In service
20	Milnes	1899	Milnes [9] S3	4 × 20hp	Winter saloon	48	37' 6" 11.43m	7' 4" 2.23m	11' 0" 3.35m	In service
21	Milnes	1899	Milnes [9] S3	4 × 20hp	Winter saloon	48	37' 6" 11.43m	7' 4" 2.23m	11' 0" 3.35m	In service
22	Milnes	1899	Milnes [9] S3	4 × 20hp	Winter saloon	48	37' 6" 11.43m	7' 4" 2.23m	11' 0" 3.35m	Burnt out 1990 [10]

Continued overleaf

Manx Electric Railway Fleet List—*continued*

Power Cars

Car No.	Built by	Year	Bogies by	Motors	Car Type	No. of Seats	Length	Width	Height	Notes
22	McArds (Port Erin)	1992	Brill 27Cx	4 × 25hp	Winter saloon	48	37' 6" 11.43m	7' 4" 2.23m	11' 0" 3.35m	Replacement [10]
23	IOMT&EP Co. Ltd	1900			Locomotive		20' 6" 6.24m	7' 6" 2.28m	11' 0" 3.35m	Damaged in a collision 1914 [11]
23	MER Co. Ltd	1925			Locomotive		34' 6" 10.51m	6' 3" 1.90m	10' 0" 3.04m	Rebuild from above Serviceable [11]
24	Milnes	1898	Brush Type D	4 × 25hp (1903)	Cross bench open	56	35' 5" 10.79m	6' 3" 1.90m	10' 6" 3.20m	Destroyed Laxey 1930 [12]
25	Milnes	1898	Brush Type D	4 × 25hp (1903)	Cross bench open	56	35' 5" 10.79m	6' 3" 1.90m	10' 6" 3.20m	In service [12]
26	Milnes	1898	Brush Type D	4 × 25hp (1903)	Cross bench open	56	35' 5" 10.79m	6' 3" 1.90m	10' 6" 3.20m	In service [12]
27	Milnes	1898	Brush Type D	4 × 25hp (1903)	Cross bench open	56	35' 5" 10.79m	6' 3" 1.90m	10' 6" 3.20m	In service [12]
28	ER&TCW Ltd	1904	Brill [13] 27Cx	4 × 25hp	Cross bench open	56	35' 0" 10.66m	6' 3" 1.90m	10' 6" 3.20m	Stored
29	ER&TCW Ltd	1904	Brill [13] 27Cx	4 × 25hp	Cross bench open	56	35' 0" 10.66m	6' 3" 1.90m	10' 6" 3.20m	Possibly serviceable
30	ER&TCW Ltd	1904	Brill [13] 27Cx	4 × 25hp	Cross bench open	56	35' 0" 10.66m	6' 3" 1.90m	10' 6" 3.20m	Stored
31	ER&TCW Ltd	1904	Brill [13] 27Cx	4 × 25hp	Cross bench open	56	35' 0" 10.66m	6' 3" 1.90m	10' 6" 3.20m	Serviceable
32	UEC	1906	Brill 27Cx	4 × 27½hp	Cross bench open	56	35' 0" 10.66m	6' 3" 1.90m	10' 6" 3.20m	In service
33	UEC	1906	Brill 27Cx	4 × 27½hp	Cross bench open	56	35' 0" 10.66m	6' 3" 1.90m	10' 6" 3.20m	In service

Notes:

[1] Brush type D bogies and 4 × 25hp motors fitted in place of Milnes S3 in 1903.
[2] Oldest electric tramcar in the world still running on its original line.
[3] Received Milnes S3 bogies and 4 × 20hp motors from car 16 in 1899.
[4] Converted to freight trailer No. 26 in 1918, currently in Ramsey Museum pending full restoration to 1895 condition.
[5] Converted to motor freight car No. 21 in 1904, bodywork removed in 1926, chassis still in use as a flat wagon.
[6] Converted to motor cattle car No. 22 in 1903, broken up in 1927.
[7] Converted to freight trailer No. 23 in 1918, renumbered 22 in March 1927 to allow locomotive No. 23 to have its number back! Finally broken up at Dhoon Quarry sidings sometime between 1955 and 1959, during a tidying up campaign.
[8] Received Milnes S3 bogies and 2 × 25hp motors from car No. 4 in 1899; Brush type D bogies and 4 × 25hp motors fitted in 1903.
[9] Nos 19–22 received Brill 27Cx bogies and 4 × 25hp motors from car Nos 29, 28, 30 and 31 respectively in 1904.
[10] Original car No. 22 destroyed by fire at Derby Castle depot during the night of 30th September 1990. New body built by McArds of Port Erin and electrical repairs carried out by staff at Derby Castle. Re-entered service in May 1992.
[11] The original No. 23 borrowed bogies from No. 17 when required, borrowing from No. 33 after rebuilding. Disused after 1944, stored at Laxey until superficially restored in 1978; restored to working order in 1983/4 and for 1993. Named *Dr R. Preston Hendry* at Derby Castle on 25th May 1992.
[12] Entered service in 1898 as trailers Nos 40–43, motorised by the MER in 1903.
[13] Nos 28–31 received Milnes S3 bogies and 4 × 20hp motors from car Nos 20, 19, 21 and 22 respectively in 1904.

Trailer Cars

13 trailers (Nos 40-48, 55, 56, 61 and 62) are still in regular service and 11 more (Nos 13, 36, 37, 49, 50, 53, 54 and 57–60) are serviceable or stored.

Snaefell Mountain Railway Fleet List

Car No.	Built by	Year	Bogies by	Motors	Car Type	No. of Seats	Length	Width	Height	Notes
1	G.F. Milnes	1895	Milnes Special	4 × 25hp	Vestibuled Saloon	46	35' 7½" 10.85m	7' 3" 2.20m	10' 4½" 3.16m	New bogies and control equipment 1977
2	G.F. Milnes	1895	Milnes Special	4 × 25hp	Vestibuled Saloon	46	35' 7½" 10.85m	7' 3" 2.20m	10' 4½" 3.16m	New bogies and control equipment 1977/78
3	G.F. Milnes	1895	Milnes Special	4 × 25hp	Vestibuled Saloon	46	35' 7½" 10.85m	7' 3" 2.20m	10' 4½" 3.16m	New bogies and control equipment 1977/78
4	G.F. Milnes	1895	Milnes Special	4 × 25hp	Vestibuled Saloon	46	35' 7½" 10.85m	7' 3" 2.20m	10' 4½" 3.16m	New bogies and control equipment 1978/79
5	G.F. Milnes	1895	Milnes Special	4 × 25hp	Vestibuled Saloon	46	35' 7½" 10.85m	7' 3" 2.20m	10' 4½" 3.16m	Destroyed by fire at the summit 16/8/1970
5	H.D. Kinnan (Ramsey)	1971	Milnes Special	4 × 25hp	Vestibuled Saloon	48	35' 7½" 10.85m	7' 3" 2.20m	10' 4½" 3.16m	Replacement on original chassis. New bogies and equipment 1978/79.
6	G.F. Milnes	1895	Milnes Special	4 × 25hp	Vestibuled Saloon	46	35' 7½" 10.85m	7' 3" 2.20m	10' 4½" 3.16m	New bogies and control equipment 1978/79
7	G.F. Milnes	1896	—	—	Freight car	Nil	27' 8" 8.43m	6' 6¾" 2.00m	10' 0" 3.04m	Unpowered, bogies borrowed from No. 5

Notes:

Passenger Cars: As built these cars had unglazed windows; glazed sliding windows were fitted by April 1896.

The lack of ventilation, when all windows were closed in sunny but windy weather, resulted in clerestory windows being fitted during the winter of 1896/97.

The seating capacity was increased in all cars to 48 by adding two single seats, one at each end backing onto the bulkhead.

Between 1977 and 1979, all six cars were fitted with new bogies built by London Transport at their Acton works and with 4 × 61 hp traction motors and control equipment from six tramcars bought from Aachen in Germany.

Freight Car: Nicknamed 'Maria', this freight car saw little use after 1924, except during the Second World War when it was used to collect peat from the Bungalow area to relieve the fuel shortages on the Island. It existed in the late 1970s but had disappeared by 1981, presumably broken up.

Groudle Glen Railway Locomotive List

Steam Locomotives

No.	Name	Builder	Date	Builder's No.	Type	Coupled Wheel Diameter	Pony Wheel Diameter	Cylinders	Weight	Current Status
—	Sea Lion	W.G. Bagnall	1896	1484	2-4-0T	1' 2"	0' 9"	$4\frac{1}{2}" \times 7\frac{1}{2}"$	4t 12cwt	In service
—	Polar Bear	W.G. Bagnall	1905	1781	2-4-0T	1' $3\frac{1}{4}"$	0' 9"	$5" \times 7\frac{1}{2}"$	5t 10cwt	Preserved at Amberley, West Sussex

Diesel and Electric Locomotives

No.	Name	Builder	Date	Builder's No.	Length	Width	Height	Wheel Arrangement	Weight	Current Status
—	Sea Lion	BEV	1921	312 or 313	9' $7\frac{1}{4}"$ 2.93m	4' 2" 1.27m	6' $4\frac{1}{4}"$ 1.94m	4-wheel	4 tons	Withdrawn 1927 Broken up [1]
—	Polar Bear	BEV	1921	312 or 313	9' $7\frac{1}{4}"$ 2.93m	4' 2" 1.27m	6' $4\frac{1}{4}"$ 1.94m	4-wheel	4 tons	Withdrawn 1927 Broken up [2]
1	Dolphin	Hunslet	1952	4394	8' 11" 2.71m	3' 5" 1.04m	5' 0" 1.52m	4-wheel	5 tons	In service
2	Walrus	Hunslet	1952	4395	8' 11" 2.71m	3' 5" 1.04m	5' 0" 1.52m	4-wheel	5 tons	In service

A diesel locomotive is known to have worked on the line in the 1950s but no further details are available.

Notes:

[1] Pony wheels were fitted at both ends about 1923. These were removed in 1925, when the batteries were transferred to a separate trolley.
[2] *Polar Bear* is known to have derailed and rolled to the bottom of the Glen (date unknown) and may never have re-entered service. Note, [1] may also apply to this locomotive.
BEV = British Electric Vehicles of Southport.

Douglas Horse Tramway Fleet List

No.	Builder	Date Built	Type	Length	Floor Width	Width Overall	Height	Notes
1	Starbuck	1876	S/D Saloon	Dimensions unknown				Converted to double deck 1884/85 Withdrawn about 1900
1	Milnes Voss	1913	S/D Saloon	24' 8" 7.51m	6' $4\frac{1}{2}"$ 1.94m	6' 7" 2.0m	9' 11" 3.02m	In service 1992
2	Starbuck	1876	Double Deck	Dimensions unknown				Broken up 1948/49
3	Starbuck	1876	Double Deck	Dimensions unknown				Broken up 1948/49
4	Starbuck	1882	Double Deck	Dimensions unknown				Broken up 1948/49
5	Starbuck	1883	Double Deck	Dimensions unknown				Broken up 1949

No.	Maker	Date	Type	Length	Width	Height	Overall height	Notes
6	Starbuck	1883	Double Deck	Dimensions unknown				Broken up 1949
7	Starbuck	1884	Double Deck	Dimensions unknown				Broken up 1924
8	Starbuck	1884	Double Deck	Dimensions unknown				Broken up 1949
9	Starbuck	1885	Toastrack	22' 9" 6.93m	5' 4" 1.62m	6' 8" 2.03m	9' 6" 2.89m	Broken up 1952
10	Starbuck	1885	Toastrack	24' 0" 7.31m	5' 4" 1.62m	6' 8" 2.03m	9' 6" 2.89m	Broken up 1983
11	Starbuck	1886	Toastrack	22' 9" 6.93m	5' 4" 1.62m	6' 8" 2.03m	9' 6" 2.89m	Withdrawn 1978 Preserved at Ramsey
12	G. F. Milnes	1888	Toastrack	22' 9" 6.93m	5' 4" 1.62m	6' 8" 2.03m	9' 6" 2.89m	In service 1992
13	MRCW Saltley	1883	Double Deck	22' 6" 6.85m	—	6' 0" 1.82m	10' 8" 3.25m	Acquired from South Shields 1887 Renumbered 14 in 1908. Last used 1939 Now preserved [1]
14	MRCW Saltley	1883	Double Deck	Dimensions unknown				Acquired from South Shields 1887 Destroyed by rock fall at depot 1908
15	MRCW Saltley	1883	Double Deck	22' 6" 6.85m	—	6' 0" 1.82m	10' 8" 3.25m	Acquired from South Shields 1887 Last used 1939. Broken up 1949
16	MRCW Saltley	1883	Double Deck	Dimensions unknown				Acquired from South Shields 1887 Broken up 1915
17	MRCW Saltley	1883	Double Deck	Dimensions unknown				Acquired from South Shields 1887 Converted to single deck 1903 Broken up 1917
18	MRCW Saltley	1883	Double Deck	Dimensions unknown				Acquired from South Shields 1887 Converted to single deck 1903–06 and back to double deck 1988/89. In service 1992
19	G.F. Milnes	1889	Toastrack	22' 9" 6.93m	5' 4" 1.62m	6' 8" 2.03m	9' 6" 2.89m	Withdrawn 1949 Broken up 1952
20	G.F. Milnes	1889	Toastrack	22' 9" 6.93m	5' 4" 1.62m	6' 8" 2.03m	9' 6" 2.89m	Withdrawn 1949 Broken up 1952
21	G.F. Milnes	1890	Toastrack	24' 8" 7.51m	5' 4" 1.62m	6' 10½" 2.09m		In service 1992
22	G.F. Milnes	1890	Toastrack	22' 5" 6.83m	5' 5" 1.65m	6' 10" 2.08m	9' 9" 2.97m	Withdrawn 1978 Converted for use as a 'Tram shop' at Derby Castle depot
23	G.F. Milnes	1891	Toastrack	22' 8" 6.90m	5' 5" 1.65m	6' 11" 2.10m		Broken up 1952
24	G.F. Milnes	1891	Toastrack	22' 8" 6.90m	5' 5" 1.65m	6' 11" 2.10m		Broken up 1952
25	G.F. Milnes	1891	Toastrack	22' 8" 6.90m	5' 5" 1.65m	6' 11" 2.10m		Broken up 1952
26	G.F. Milnes	1891	Toastrack	22' 8" 6.90m	5' 5" 1.65m	6' 11" 2.10m		Broken up 1974
27	G.F. Milnes	1892	Winter Saloon	24' 5" 7.44m	5' 10" 1.77m	6' 6" 1.98m	9' 2" 2.79m	In service 1992 [2]
28	G.F. Milnes	1892	Winter Saloon	24' 5" 7.44m	5' 10" 1.77m	6' 6" 1.98m	9' 2" 2.79m	In service 1992 [2]
29	G.F. Milnes	1892	Winter Saloon	24' 5" 7.44m	5' 10" 1.77m	6' 6" 1.98m	9' 2" 2.79m	In service 1992 [2]
30	G.F. Milnes	1894	Toastrack	22' 6" 6.85m	5' 5" 1.65m	6' 10" 2.08m		Broken up 1952
31	G.F. Milnes	1894	Toastrack	22' 6" 6.85m	5' 5" 1.65m	6' 10" 2.08m		Broken up 1987

Continued overleaf

Douglas Horse Tramway Fleet List—*continued*

No.	Builder	Date Built	Type	Length	Floor Width	Width Overall	Height	Notes
32	G.F. Milnes	1896	Toastrack	21' 8" 6.60m	5' 4" 1.62m	6' 10" 2.08m	8' 7" 2.61m	In service 1992
33	G.F. Milnes	1896	Toastrack	21' 8" 6.60m	5' 4" 1.62m	6' 10" 2.08m	8' 7" 2.61m	In service 1992
34	G.F. Milnes	1896	Toastrack	21' 8" 6.60m	5' 4" 1.62m	6' 10" 2.08m	8' 7" 2.61m	In service 1992
35	G.F. Milnes	1896	Toastrack	21' 8" 6.60m	5' 4" 1.62m	6' 10" 2.08m	8' 7" 2.61m	In service 1992
36	G.F. Milnes	1896	Toastrack	24' 11" 7.59m	5' 4" 1.62m	6' 10" 2.08m	8' 7" 2.61m	In service 1992
37	G.F. Milnes	1896	Toastrack	21' 8" 6.60m	5' 4" 1.62m	6' 10" 2.08m	8' 7" 2.61m	In service 1992
38	G.F. Milnes	1902	Toastrack	24' 5" 7.44m	?	?		In service 1992
39	G.F. Milnes	1902	Toastrack	23' 0" 7.01m	5' 5" 1.65m	6' 11" 2.10m		Stored
40	G.F. Milnes	1902	Toastrack	24' 5" 7.44m	5' 4" 1.62m	6' 10" 2.08m		Broken up 1991/2
41	Milnes Voss	1905	Toastrack	23' 3" 7.08m	5' 6" 1.67m	7' 1" 2.16m		Renumbered 10 in 1985 Withdrawn 1988
42	Milnes Voss	1905	Toastrack	24' 8" 7.51m	5' 5" 1.65m	7' 0" 2.13m		In service 1992
43	United Electric Car Company	1907	Toastrack	24' 6" 7.46m	5' 5" 1.65m	6' 11" 2.10m	8' 10" 2.69m	Temporarily renumbered 16 1987/88 Now stored
44	United Electric Car Company	1907	Toastrack	24' 6" 7.46m	5' 5" 1.65m	6' 11" 2.10m	8' 10" 2.69m	In service 1992 [3]
45	Milnes Voss	1908	Toastrack	25' 0" 7.62m	5' 6" 1.67m	7' 0" 2.13m	8' 6" 2.59m	Stored
46	Milnes Voss	1909	Toastrack	24' 11" 7.59m	5' 5" 1.65m	6' 11" 2.10m	8' 11" 2.71m	Withdrawn 1987 Preserved in England [4]
47	Milnes Voss	1911	Toastrack	25' 1" 7.64m	5' 5" 1.65m	7' 0" 2.13m	8' 9" 2.66m	Withdrawn 1978 Preserved at Ramsey
48	Vulcan Motor	1935	Saloon/ Toastrack Convertible	25' 6" 7.77m	5' 7" 1.70m	6' 11" 2.10m	8' 6" 2.59m	Withdrawn 1980 Broken up 1982
49	Vulcan Motor	1935	Saloon/ Toastrack Convertible	25' 6" 7.77m	5' 7" 1.70m	6' 11" 2.10m	8' 6" 2.59m	Withdrawn 1980 Preserved at Ramsey
50	Vulcan Motor	1935	Saloon/ Toastrack Convertible	25' 6" 7.77m	5' 7" 1.70m	6' 11" 2.10m	8' 6" 2.59m	Withdrawn 1980 Broken up 1982

Notes:
[1] Left the Island in 1955, later displayed at the Museum of British Transport, Clapham, returning for the centenary celebrations of 1976 and now on display in the Manx Museum, Douglas.
[2] Originally built without platform vestibules which were added in 1895.
[3] The Royal car, having been used on several occasions to convey members of the Royal Family.
[4] Preserved at the ferry terminal, Woodside, Birkenhead.

General note:
Many minor alterations have taken place over the years. Several cars have been lengthened and have had seating alterations; the dimensions shown here are the latest known in each case.

142

Steam Railway Locomotive List

Steam Locomotives

No.	Name	Builder	Date	Builder's No.	Type	Coupled Wheel Diameter	Pony Wheel Diameter	Cylinders	Weight	Current Status
1	Sutherland	Beyer, Peacock	1873	1253	2-4-0T	3' 9"	2' 0"	11" × 18"	19 tons	Museum Port Erin
2	Derby	Beyer, Peacock	1873	1254	2-4-0T	3' 9"	2' 0"	11" × 18"	19 tons	Dismantled 1951
3	Pender	Beyer, Peacock	1873	1255	2-4-0T	3' 9"	2' 0"	11" × 18"	19 tons	Sold (off Island) 1978 [1]
4	Loch	Beyer, Peacock	1874	1416	2-4-0T	3' 9"	2' 0"	11" × 18"	19 tons	In service
5	Mona	Beyer, Peacock	1874	1417	2-4-0T	3' 9"	2' 0"	11" × 18"	19 tons	In store Douglas [2]
6	Peveril	Beyer, Peacock	1875	1524	2-4-0T	3' 9"	2' 0"	11" × 18"	19 tons	In store Douglas
7	Tynwald	Beyer, Peacock	1880	2038	2-4-0T	3' 9"	2' 0"	11" × 18"	19 tons	Dismantled 1946 [2]
8	Fenella	Beyer, Peacock	1894	3610	2-4-0T	3' 9"	2' 0"	11" × 18"	19 tons	In store Douglas [2]
9	Douglas	Beyer, Peacock	1896	3815	2-4-0T	3' 9"	2' 0"	11" × 18"	19 tons	In store Port Erin [2]
10	G.H. Wood	Beyer, Peacock	1905	4662	2-4-0T	3' 9"	2' 0"	11" × 18"	21 tons	In service [3]
11	Maitland	Beyer, Peacock	1905	4663	2-4-0T	3' 9"	2' 0"	11" × 18"	21 tons	In service
12	Hutchinson	Beyer, Peacock	1908	5126	2-4-0T	3' 9"	2' 0"	11" × 18"	21 tons	In service
13	Kissack	Beyer, Peacock	1910	5382	2-4-0T	3' 9"	2' 0"	11" × 18"	21 tons	In service [4]
14	Thornhill	Beyer, Peacock	1880	2028	2-4-0T	3' 9"	2' 0"	11" × 18"	19 tons	Withdrawn 1958 Preserved privately on IOM [5]
15	Caledonia	Dübs & Co.	1885	2178	0-6-0T	3' 3"	—	13" × 20"	27 tons	Museum Port Erin [6]
16	Mannin	Beyer, Peacock	1926	6296	2-4-0T	3' 9"	2' 0"	12" × 18"	23t 6½cwt	Museum Port Erin
	Ramsey	Sharp, Stewart	1879	2885	2-4-0T	3' 9"	2' 9"	11" × 18"	17t 15cwt	Scrapped 1923 at Douglas [7]
	Northern	Sharp, Stewart	1879	2886	2-4-0T	3' 9"	2' 9"	11" × 18"	17t 15cwt	Scrapped 1912 at Douglas [8]

Diesel Locomotive and Railcars

No.	Builder	Date	Builder's No.	Length	Width	Height	Wheel Arrangement	Weight	Current Status
17	Schoema	1958	2175		7' 2½" 2.20m	8' 10¼" 2.70m	4-wheel	21 tons	In service [9]
19	Walker/GNR(I)	1949	—	41' 2¾" 12.56m	7' 6" 2.28m	9' 11" 3.02m	Railcar		Serviceable [10]
20	Walker/GNR(I)	1950	—	41' 2¾" 12.56m	7' 6" 2.28m	9' 11" 3.02m	Railcar		Serviceable [10]

Notes:

[1] Displayed at the Greater Manchester Museum of Science and Industry as a sectionalised exhibit.
[2] Property of Isle of Man Railway Society. The frames of No. 7 are on display in the goods yard at Castletown. It is hoped to return No. 8 to service during 1993.
[3] Return to service planned for 1993 after 16 years awaiting a new boiler.
[4] One of the serviceable fleet but likely to be out of service for four to five years while a complete rebuild takes place and a new boiler is obtained. Original boiler now on No. 10.
[5] Ex-Manx Northern Railway No. 3.
[6] Ex-Manx Northern Railway No. 4. Possibility of return to service for Snaefell centenary in 1995, as this locomotive was used during the construction of the line.
[7] Ex-Manx Northern Railway No. 1.
[8] Ex-Manx Northern Railway No. 2.
[9] Purchased from Germany in 1992 for works duties to avoid the cost of raising steam for works trains.
[10] Ex-County Donegal railcars transferred to Isle of Man in 1961. In working condition but in need of attention before use on passenger services again.

14

Manx Tail Piece

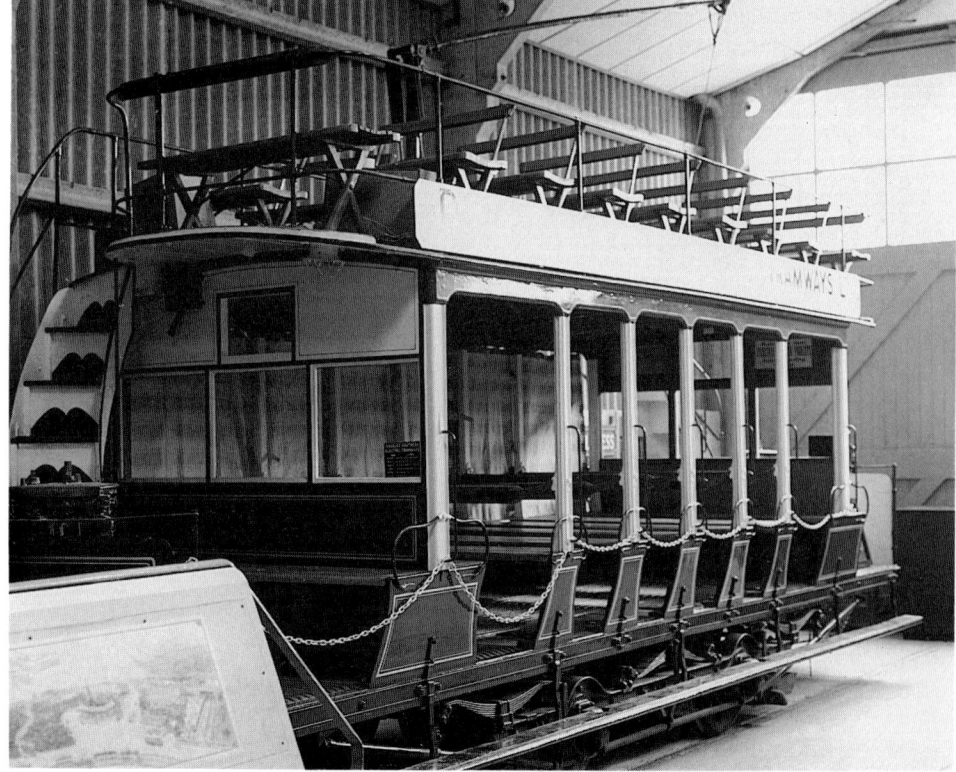

Laxey Mines. Laxey was once a lead and silver mining town. A small railway was built within the mines, at first operated by horses and later by two tiny 16in gauge 0-4-0 steam locomotives. Built by Stephen Lewin with builder's numbers 684 and 685, they were named *Ant* and *Bee*. Following the closure of the mines and therefore the railway, the locomotives were sold to a Douglas scrap merchant in 1935. As part of a major scheme to re-open the mine as a tourist attraction, a short section of the railway has been made safe and this picture shows a wagon loaded with ore standing on the restored line. The electric lighting is a recent addition!

Douglas Southern Tramway. One other surviving piece of Isle of Man tramway equipment is Douglas Southern tramcar No. 1. Built by Brush in 1896 on a 'Lord Baltimore', USA chassis, it is 28ft 4in (8.63m) long, 7ft 4in (2.23m) wide and 14ft 4in (4.36m) to the top of the trolley standard. It was rescued from the depot site at Little Ness during 22/23rd June 1951 and, after being exhibited at Clapham Museum, was moved to the Crich Tramway Museum in Derbyshire, where it has been restored to working order, as seen here. Unusually for the Isle of Man, this line was built to the British standard gauge of 4ft 8½in.